Roberto Esposito

SUNY series in Contemporary Italian Philosophy

Silvia Benso and Brian Schroeder, editors

Roberto Esposito

Biopolitics and Philosophy

EDITED BY

Inna Viriasova
AND Antonio Calcagno

Published by State University of New York Press, Albany

Printed in the United States of America

For information, contact State University of New York Press, Albany, NY
www.sunypress.edu

Library of Congress Cataloging-in-Publication Data

Names: Viriasova, Inna, editor.
Title: Roberto Esposito : biopolitics and philosophy / edited by Inna Viriasova and Antonio Calcagno.
Other titles: Roberto Esposito (State University of New York Press)
Description: Albany, NY : State University of New York Press, 2018. | Series: SUNY series in contemporary Italian philosophy | Includes bibliographical references and index.
Identifiers: LCCN 2017034890 | ISBN 9781438470351 (hardcover : alk. paper) | ISBN 9781438470368 (ebook)
Subjects: LCSH: Biopolitics—Philosophy. | Political science—Philosophy. | International relations—Philosophy. | Esposito, Roberto, 1950–
Classification: LCC JA80 .R638 2018 | DDC 320.01—dc23
LC record available at https://lccn.loc.gov/2017034890

10 9 8 7 6 5 4 3 2 1

Contents

Abbreviations vii

Introduction: Thinking with Roberto Esposito ix
Inna Viriasova

Part I: Reading Esposito

1. The Presence and Absence of Origin: Roberto Esposito's Early Interpretation of Giambattista Vico 3
Alexander U. Bertland

2. Thinking Community 27
Diane Enns

3. Debt and the Proper in Agamben and Esposito 47
Greg Bird

4. Against the Conspiracy. Revisiting Life's Vertigo: On Roberto Esposito's *Terza persona* and *Da fuori* 65
Alberto Moreiras

5. Fold of Life: Roberto Esposito on "the Living Person" and Animistic Personhood 101
Inna Viriasova

6. The Failure of the Political Concept of the Person? A Foucaultian-Arendtian Response to Roberto Esposito 127
Antonio Calcagno

7. Person and *Munus* in the Thought of Roberto Esposito 143
Jonathan Short

8. From Biopolitics to Political Animism:
Roberto Esposito's Things 161
Federico Luisetti

Part II: Applying Esposito

9. The Impolitical Dimension in Jorge Luis Borges's Literature:
A Gaze on the Impossible for Politics through
Roberto Esposito's Thought 179
Federico Fridman

10. Internalities of International Relations and the Politics of
Externalities: Affirming the Impossibility of IR with
Roberto Esposito 201
Mark F. N. Franke

11. Living Through Catastrophe: Warring Immunities,
Dramatization and Counter-Actualization in
Wajdi Mouawad's *Scorched* 219
Geoffrey Whitehall

12. Becoming Normative: Law, Life, and the Possibility of an
Affirmative Biopolitics 241
Patrick Hanafin

Contributors 259

Index 263

Abbreviations

Esposito's Works in Italian

(*CIM*) *Categorie dell'impolitico* (Bologna: Il Mulino, 1999).

(DF) *Da fuori. Una filosofia per l'Europa* (Torino: Einaudi, 2016).

(*DP*) *Dieci pensieri sulla politica* (Bologna: Il Mulino, 2011).

(*II*) *Dall'impolitico all'impersonale: conversazioni filosofiche* (Milano: Mimesis, 2012).

(MV) *La politica e la storia. Machiavelli e Vico* (Napoli: Liguori, 1980).

(*PC*) *Le persone e le cose* (Torino: Einaudi, 2014).

(*TPN*) *Terza persona. Politica della vita e filosofia dell'impersonale* (Torino: Einaudi, 2007).

(VR) *Vico e Rousseau e il moderno Stato borghese* (Bari: De Donato, 1976).

Esposito's Works in English Translation

(BS) *Bios: Biopolitics and Philosophy*, trans. Timothy Campbell (Minneapolis, MN: Minnesota University Press, 2008).

(CI) *Categories of the Impolitical*, trans. Connal Parsley (New York: Fordham University Press, 2015).

(*CM*) *Communitas: The Origin and Destiny of Community*, trans. Timothy Campbell (Stanford, CA: Stanford University Press, 2004).

(*IM*) *Immunitas: The Protection and Negation of Life*, trans. Zakiya Hanafi (Cambridge, UK; Malden, MA: Polity Press, 2011).

(*LT*) *Living Thought: The Origins and Actuality of Italian Philosophy*, trans. Zakiya Hanafi (Stanford, CA: Stanford University Press, 2012).

(*PT*) *Persons and Things: From the Body's Point of View*, trans. Zakiya Hanafi (Cambridge, UK; Malden, MA: Polity Press, 2015).

(*TP*) *Third Person: Politics of Life and Philosophy of the Impersonal*, trans. Zakiya Hanafi (Cambridge, UK; Malden, MA: Polity Press, 2012).

(*TTP*) *Terms of the Political: Community, Immunity, Biopolitics*, trans. R. Noel Welch (New York: Fordham University Press, 2013).

(*TW*) *Two: The Machine of Political Theology and the Place of Thought*, trans. Zakiya Hanafi (New York: Fordham University Press, 2015).

Introduction

Thinking with Roberto Esposito

Inna Viriasova

This book is dedicated to the work of the contemporary Italian philosopher Roberto Esposito. He is currently professor of philosophy at the Scuola Normale Superiore at Pisa, Italy. Esposito appeared on the philosophical scene in the 1970s with contributions to the history of political ideas, including major publications on Machiavelli, Vico, and Rousseau; he has gradually become one of the most recognized names in contemporary debates in political philosophy through his extensive work on biopolitics, community, immunity, and the person.

Over the past two decades we have witnessed a growing interest in Esposito's work, and his thought continues to have an impact on scholarship in many disciplines, including politics, sociology, literature, and philosophy. As his work becomes better known in the English-speaking world, more attention is being paid specifically to his intervention in the field of biopolitical theory opened up by Michel Foucault. Esposito's particular contribution in this regard lies in his development of the project of affirmative biopolitics, which explores the ways in which life, despite its unavoidable politicization in modernity, can nevertheless regain and maintain its creative potential. Due to its innovative and comprehensive nature, Esposito's writing has already garnered special attention in many leading academic journals, and a number of books are being published that focus on his thought vis-à-vis Italian philosophy. A number of conferences have also been held in Europe and North America, over the past decade, which focused on Esposito's work.[1]

The primary intention of the present collection of essays is to offer readers a comprehensive introduction to and critical explanation of Esposito's political thought and the key concepts that he has developed up until now. In this regard, our goal is twofold: in addition to exploring some of Esposito's major conceptual contributions, this book offers a number of their creative applications to various fields of inquiry, such as literary studies, international relations, politics, and law. To date, while there exist a few monographs in English that introduce certain aspects of Esposito's thought, this is the first volume to provide a comprehensive analysis of his core concepts while also demonstrating some practical uses of his ideas. We bring together a number of leading international scholars on the philosopher's work in order to explore and critique both his early and more recent contributions in political philosophy. The contributors address particular aspects of his growing corpus such as the impolitical, community, immunity, the impersonal, affirmative biopolitics, justice, life, the third person, and the body, as well as Esposito's reading and interpretation of classical political thinkers, including Hobbes, Machiavelli, Vico, and Kant. The reader is thus exposed to not only the most familiar and most recent elements of Esposito's oeuvre, but also to its earlier conceptual developments and contributions to the study of the history of political ideas. While Esposito's work is primarily indebted to Western philosophical tradition, a number of essays in this volume bring Esposito's ideas into a welcome conversation with non-Western and indigenous perspectives, exploring avenues for their creative collaboration and synthesis.

Several essays in this volume employ Esposito's philosophical ideas for thinking through some pertinent issues in international relations, postcolonialism, literature, science, technology, and philosophical and artistic practice, bringing Esposito into dialogue with important social-political concerns. This is a propitious time for such an endeavor considering that many of the issues that Esposito raises in his work are becoming increasingly prominent in everyday life, both at the local level and on the global scale, including environmental degradation, migration and forced displacement, war on terrorism, hybrid wars, processes of decolonization, and the growing violations of human rights. The ongoing work of translation of Esposito's early and recent works into English also makes it a good time to put multiple interpretations of his work into productive conversation with already established scholarship in the history of ideas, biopolitics, community, immunity, and personhood.

Each of the following essays draws out its singular trajectory of engagement with Esposito's thought, furthering our understanding of political being, action, personhood, life, and the meaning of the common. Each essay, as it engages the foundational concepts of Esposito's philosophy, is ultimately an exercise of thinking "with" and even "after" Esposito. Enriched by the intervention of his thought, the authors of this collection invite the reader to partake in the project of discovering how to think and live differently in a time when the negative aspects of biopolitics are progressively enclosing societies in vicious cycles of conflict, death, abuses of fundamental liberties, and environmental exploitation.

Central Themes of Esposito's Work

The chapters in this book cover the central themes that have emerged in Esposito's work over the past several decades. Even though Esposito's name has entered the stage of English-speaking scholarship with the publication of translations of *Communitas* (2004) and *Bios* (2008), associating his name primarily with the field of the post-Foucaultian biopolitical theory, his philosophical corpus, dating back to the 1970s, has in no way been limited to the question of biopolitics. Most of his early works are dedicated to critical readings of the foundational texts in the history of political philosophy, and his books on Vico, Rousseau, Machiavelli, Descartes, and Spinoza especially stand out in this regard.[2]

Political ontology occupies another prominent place in Esposito's early pre-biopolitical work, including questions of the origins and limits of politics in contemporary political philosophy.[3] These questions have been developed by Esposito through his category of the impolitical (*l'impolitico*), which, following Massimo Cacciari's lead[4], he explored in *Categorie dell'impolitico* (1988). This book was reprinted in Italian with a new preface in 1999, and was only recently published in English as *Categories of the Impolitical* (2015). Dissatisfied with the way in which the notion of the political has functioned in Western philosophy for the past several centuries, inasmuch as it has been dominated by theological and totalitarian tendencies, Esposito undertakes the task of rethinking the political from its limits. Focusing on features that have been traditionally excluded from or have been thought of as undermining the domain of the political (e.g. conflict, plurality, difference), Esposito views the impolitical not as a rejection of politics, but as an immanent critique of modern political regime of thought, traditionally indexed by representation, depoliticization, and Christian theology. The impolitical is the shadow of the political that challenges and undermines its established foundations, constructing the basis for a new, post-foundational politics that has become aware of its limits and finitude. The impolitical, even though Esposito no longer employs the term in his recent work, forms the foundation for his recent philosophy of the impersonal, which, as a critique of the *dispositif* or apparatus of the person, follows the logic of his impolitical critique of the political.

Many of Esposito's more recent works contribute to a distinct trend in twentieth-century continental philosophy, within French and Italian contexts in particular, that is concerned with rethinking the common. Among other prominent names in this regard are Georges Bataille, Maurice Blanchot, Hannah Arendt, Simone Weil, Jacques Derrida, Jean-Luc Nancy, and Giorgio Agamben. Esposito's approach to the problem of community has received significant attention from scholars and commentators in various disciplines. Importantly, for Esposito, the question of the common arises and evolves in conjunction with the question of the biopolitical, coming to fruition in the publication of his trilogy *Communitas: The Origin and Destiny of Community* (1998), *Immunitas: The Protection and Negation of Life* (2002) and *Bios: Biopolitics and Philosophy* (2004).

The first part of the trilogy, *Communitas,* examines the origin of community through the Roman notion of *munus*. It is also inseparably connected with Esposito's subsequent examination of the notion of immunity that hinges upon a different mode of operation of the very same *munus* that enables community. Traditionally community has been understood as a human unity centered around some shared property of all its members. Esposito deconstructs this notion of community beginning with its etymological analysis. The core of community, he suggests, is not a shared property but *munus*—a debt, a pledge, a gift to be given to the other, establishing a lack and an "obligation" as that which unites the subjects of community. Community is not a "body" comprised of separate individuals that come together in a contract; it is not, as Esposito puts it, "the subject's expansion or multiplication but its exposure to what interrupts the closing and turns it inside out" (*CM*, 7). A debt to the other is the "property" of community, and as such, community is (im)possible.

The concept developed by Esposito in conjunction with the notion of community is immunity, examined in detail in his book *Immunitas*. While community is shared *munus*, an opening of existence outside itself that "breaks down the barriers of individual identity," immunity is a way of constructing these barriers "in defensive and offensive forms, against any external element that threatens it."[5] An immunized element refuses to partake in the sharing of the debt of community; it does not owe anything to anyone, thus destroying the constitutive bond of the common consisting in the gift exchange. Such an element, in order to preserve itself, must develop immunity, i.e., a defense mechanism employed in its relationship or communication with exteriority. Immunity protects life from its communal element, from extending too far toward an insecure outside and at the same time keeps this outside from invading interiority. It is a membrane that regulates the relationship between the safe and the unsafe. Importantly, while immunity is necessary for the preservation of life, when it passes beyond a certain threshold, it threatens with destruction the very unity it seeks to protect: the closure of immunity may devastate the opening of community.

Esposito dedicated a significant part of his work to analyzing and critiquing, as well as finding ways of resolving this paradoxical aspect of the relationship between immunity and community, forming one of the major tasks of his project of "affirmative biopolitics." In Foucault's early interpretation, biopolitics emerges in modernity and consists in government of life through a network of power and knowledge relations that operate both at the level of the state apparatus and, most importantly, at the micro-level of social interactions and non-state institutions. After Foucault, there have emerged a number of important negative and positive interpretations of biopolitics. The former interpretation views biopolitics primarily as a repressive political apparatus that exercises power *over* the living populations, thus stifling life's creative and communal potential. In this regard, power performs a clear-cut immunitary function. However, immunity, Esposito argues, does not operate only negatively; on the contrary, by its very definition, it requires a relationship of opening toward "hostile" exteriority. In order for immunity to develop at all, the outside must penetrate the inside. As such, in governing life, even as politics performs a limiting immunitary or regulative role, it cannot prevent life from becoming more than it is at a given moment. In other words, politics can never fully deprive life of its power and communal potential. Immunity and community, politics and life do not exclude but fold into each other, with life triumphing even in the face of death and destruction. Biopolitics, then, designates not merely power *over* life but also necessarily power *of* life. It both negates *and* affirms life.

Esposito's project of affirmative biopolitics laid the ground for his recent work on the question of the person—one of the central *dispositifs* of political government in Western societies. In particular, the problem of a creative transcendence of existing forms and norms of political being motivates Esposito's critique of the *dispositif* of the person, which has received significant attention in recent scholarship, particularly since the publication of *Third Person: Politics of Life and Philosophy of the Impersonal* and its English translation. In this book, Esposito presents a critical genealogy of the person, exposing the mechanisms that have been for a long time responsible for the logic of the exception of any form of life that is unable to provide the credentials of personhood. In the process of the institution of human community, Esposito shows, the impersonal is traditionally confined to the domains of animality, including mere humanity, and thinghood, none of which can claim the right of belonging to the common protected by the rule of law. The proud bearer of rights in Western societies—the person—has been constituted in opposition to and, most tragically, to the detriment of "things" or that which is merely "formally" human. In light of this exposition, Esposito urges us to challenge the absolute onto-theological primacy of the person, and suggests that such a challenge, in the figure of the third person, has surfaced a number of times in twentieth-century philosophy. In this regard, he thoroughly examines the implications for his new philosophy of the impersonal of Benveniste's non-person, Kojève's animal, Jankélévitch's the Other, Levinas's face of the other, Blanchot's neuter, Foucault's outside, Deleuze's event and becoming-animal, and Simone Weil's impersonal. Esposito's genealogy of the impersonal is an interruption of the dominant theoretical narrative of the person, which results, however, not in a complete eradication of this notion but in its creative reworking through the idea of "the living person"—an alternative *dispositif* of power that replaces the logic of exception with an opening toward nonhuman exteriority.

The latest addition to Esposito's work on the person is his book *Persons and Things*, where he continues the line of critique introduced in *Third Person* regarding the constitutive opposition between persons and things. The continuing attention to this opposition, Esposito suggests, leaves "the body" out of focus and places it in an ambiguous position: covered in silence, the body has become the ground for the operation of biopolitical government. At the same time, the body, as the domain of mediation between persons and things, offers a possibility for a reworking of this exclusionary dualism into a new positive alliance between persons and things, mitigating the negative effects of the modern biopolitical regime. As I write this introduction, Esposito has turned his attention to Europe and the crisis it is undergoing. His recently published work, *Da fuori. Una filosofia per l'Europa* (*From Outside: A Philosophy for Europe*), collects various essays on the theme of Europe, its culture, politics, and history. Undoubtedly, as Europe faces challenges like Brexit, the themes of community and belonging become central to European identity.

Contributions

Following the major themes of Esposito's prolific corpus, the present volume consists of two parts. The first part loosely follows the chronology of Esposito's published works, exploring their conceptual contributions to the history of political thought as well as contemporary political philosophy. This part contains a chapter dedicated to Esposito's early work on Giambattista Vico, several contributions exploring the notions of community and immunity, the critique of the person, the concept of the impersonal, political animism, and the most recent work on persons and things. The second part offers some examples of critical engagement with Esposito's thought in such areas as literary studies, theatre, international relations, and law.

The opening essay of the volume by Alexander Bertland offers an examination of Esposito's early interpretation of Vico and explores how Esposito's approach in these early texts served as an initial proving ground for his developing philosophical approach. The latter is distinguished by the way in which, in his reading of familiar texts, Esposito aspires to show not only the flexibility of established moral norms, but also how previously hidden necessities are guiding the very historical process of normalization. Bertland suggests that Vico's *New Science* provides an arena in which Esposito deploys his philosophical approach, ultimately laying the foundation for his later, more famous genealogical investigations. Esposito's reading of Vico forces him to ask questions that other readers have overlooked, clearing a path for the connection of Vico to modern political thought. Vico's ideas have been traditionally discussed in the context of the anti-Enlightenment, i.e., in opposition to the greater Enlightenment project and specifically to that of René Descartes. Esposito shifts this context by putting Vico into conversation with Jean-Jacques Rousseau and Machiavelli while focusing on the question of the relationship of philosophy and history elaborated by these authors. Bertland points out that the focal point of

Esposito's investigation here becomes the difference in their understanding of the relationship between universal truth and particular events within the course of history. Through a re-centering of Vico's thought, Esposito argues that to understand the universal it is necessary to understand the historical existence of the particular: the universal comes to fruition out of particular institutions and not the other way around. This is what allows Vico, Esposito contends, to find and endorse courses of political action that Rousseau and Machiavelli miss because they fail to connect history and metaphysics.

Esposito's notion of community, understood as an obligation or debt of exposure to the other, is the focus of Diane Enns's critical contribution. Enns sets out to introduce Esposito's analysis of community and immunity and, most importantly, to reflect on the implications and limitations of his etymological-philosophical account by considering actual experiences of (re)building communities. Rather than insisting on the impossibility of community, as Esposito has it, Enns argues that we do in fact know what community is, we experience it in its "fragile imperfection, as we move collectively between the enjoyment of human togetherness and the risk of conflict and violence." While Esposito prioritizes lack, duty, and debt as the unifying principles of the common, Enns suggests that it is the enjoyment and pleasure of human togetherness that form the foundation of community and that provide us with the conditions necessary to deal with the ever-present risk of violence and conflict, i.e., of immunity growing out of proportion and destroying community. When the self opens to the other, as Esposito suggests, immunity and community become indistinguishable. This act of opening, Enns contends, may be performed not out of debt or duty, but rather out of "pleasure in shared pursuits and conversation," the kind of pleasure that we take in expressing and encountering difference as well as in the mutual affirmation of our selves through communication. When we consider the meaning of community as we really experience it, especially in a context of a community rebuilding itself after war, we will find that it is indeed such affective positivities as desire, pleasure, compassion, and love, rather than the negativity of lack or debt, that enable communities to emerge. Enns concludes that Esposito's emphasis on the *munus* underplays the affective dimensions of community that, especially in times of extreme violence, constitute its essential elements.

Greg Bird's essay situates Esposito in a wider philosophical context and in relation to the nascent phenomenon of "Italian Thought." Bird revisits Esposito's assessment of Italian thought, focusing on its questioning of the relationship between neutralization and conflict. More specifically, Bird's essay presents a comparative reading of Esposito's and Agamben's philosophies. He notes that while Esposito can be seen as one of the main proponents of Italian Thought, defined by affirmation, Agamben's philosophy is closer to the French, marked by the tendency toward neutralization. However, despite the differences in priorities, both thinkers share an investment in assessing the domination of the economic over the political, the biopolitical regulation of life, as well as the future of community. The latter is the core concern of Bird's essay. He examines the way in which Esposito and Agamben position the notion of community against the more traditional dialectic of alienation and appropriation, and partake, with many others, in the task of conceiving community beyond the "proper" and in terms of "sharing." Without sharing, Bird notes, appropriation remains a private activity, as in the case of the social contract, resulting in a "division without sharing." In other words, "relationships cannot remain relational if they are appropriated," and so what is understood as "proper" relationality always holds the improper within it. Both Agamben and Esposito accept this premise of the improper at the center of community that is particularly developed in their exploration of the notion of debt, to which, Bird argues, they turn in order to rethink and reassess the relationship between ontology, ethics, and politics in their thinking of community.

Alberto Moreiras's original contribution examines Esposito's critique of the *dispositif* of the person and the notion of the impersonal in *Third Person*, supplemented by a reading of Esposito's more recent books *Da fuori* and *Persons and Things*. Moreiras argues that what is at stake in Esposito's project is an "endeavor of beatitude": the rupture (for the sake of an affirmative post-subjective techno-politics) of the theoretical circle of the person given with its originary split between rational and animal, personal and impersonal life, between subject and object, spirit and body, substance and mode, thinking and being. Moreiras reminds us that Esposito's main thesis, namely, that the multiplying contemporary abuses of human rights have developed, paradoxically, because of the overwhelming ideological dominance of the *dispositif* of the person and not in spite of it, has a crucial consequence: the need for the elaboration of an alternative understanding of the person, without which no political transformation can be effective. Consequently, a question arises about the possibility of a third-person politics, of a politics guided by a different, impersonal regime of sense, or, as Moreiras puts it, a "politics as the countercommunitarian history of the neuter." At the end of his thorough analysis, Moreiras concludes that a politics of the third person is expressed by Esposito through the notion of "a biopolitics of impersonal life" that must lead toward a new, impersonal, and singular community of beatitude. On a critical note, Moreiras brings into conversation Agamben's *The Open*, which shares many of Esposito's insights, but also, he suggests, advances a more nuanced argument for the task of erasing the metaphysical separation between *homo* and *persona*, between animal and person.

Inna Viriasova's essay focuses on the notion of "the living person," which Esposito introduces in *Third Person*. Esposito's exploration of this notion is particularly inspired by Foucault and Deleuze, for whom the form assumed by the third person is *life*. The living person is a fold within the person, situated at the limit of the personal; it is "the person open to what has never been before." Viriasova argues that the opening, which the notion of the living person performs, is twofold. First, from a Foucaultian perspective, it is indicative of the ever-present emergence of new forms of living resistance that traverse the *dispositifs* of modern power/knowledge, and thus enable an affirmative biopolitics. Second, Esposito's reading of Deleuze on "*a* life" and "becoming-animal" implies a different opening that has a potential to traverse not only the repressive biopolitical regime of modernity, but any form of Western politics whatever. In this respect, the living person is a figure of the immanence and the indivisibility of life or an alternative *dispositif* that produces real effects at the level of life itself, prior to its problematic relationship with politics and law. The living person, Viriasova contends, situates Esposito's philosophy within a different ontological horizon that, under the general name of "new animism," has recently become an important tool of decolonization of thought in the humanities and social sciences (a similar argument is further developed in Federico Luisetti's contribution). As a result, we may further enrich Esposito's notion of the living person by turning to the animist conceptions and performances of personhood found in certain indigenous societies.

Antonio Calcagno's essay offers a critical response to Esposito's philosophy of the third person. The concept of the person, while it has been mobilized to advocate for greater rights, freedoms and protections for marginalized groups, has also been ignored and abused. Calcagno notes that Esposito's rejection of the person is motivated by the fact that the person is prone to manipulation by political powers for the sake of managing and controlling populations. Esposito argues that since the person cannot resist these machinations, it is a weak concept that can do very little to affirm the dignity and inviolability of human lives in the face of violence. Furthermore, the person is inscribed within the Western political system, dominated by discourses of security and survival, which treat the individual as a biological entity and not as a person. Last, the massive scale of recent global change turns personal individuation into a non-viable unit of political collectivity. In order to defend the concept of the person, Calcagno suggests that Esposito's critique is not completely justified. He argues that Esposito's critique relies on the discourse of law, politics, science, and technology, and ignores the impact of moral, psychological, and philosophical discourses. Employing the assistance of Arendt and Foucault, Calcagno expands the discussion of personhood by including in it a moral dimension, as well as the notion of care of the self and personal identity, all of which help overcome Esposito's charge of the political non-viability of the person.

Jonathan Short's contribution addresses the critique that Esposito's notion of the impersonal is politically indeterminate or insufficient, and argues that it can, on the contrary, be mobilized to challenge neo-liberal governmentality. First, reading Esposito alongside Foucault's and Timothy Campbell's recent work on liberal biopolitics, Short suggests that from Esposito's diagnosis of the person we can extrapolate an assessment of the way the person has become a major device of neoliberal immunization. Furthermore, in order to address a particular strand of critical response to Esposito's recent work on the impersonal as an anti-political concept, Short affirms the political relevance of the impersonal by linking it to Esposito's earlier work on community. In doing so, the impersonal is revealed as a conception of right that is not predicated on the status of the person. Instead of immunizing each from all, it connects the unique and singular aspect of each human being with what is common to all—the *munus* or an obligation, corresponding to "the rights of the entire human community." In the same manner, the gift of life itself, which is impersonal since it cannot be appropriated by any living being, constitutes a radical impropriety or a lack through which community lives in us and thus obliges us toward it. In order to develop the political import of the idea of the impolitical *munus*, Short brings it into dialogue with Jacques Rancière's notion of dissensus, providing the basis for thinking the common against the neoliberal apparatus of immunity, which turns subjects into "enterprise selves" driven by the need of competitive adaptation that destroys community. Short argues that dissensus on the basis of the *munus* as an equality of obligation may serve as the political potential or "ground zero" of claims to membership in a community.

The essay by Federico Luisetti is dedicated to Esposito's book *Persons and Things*, which not only continues the trajectory of his prolific critique, but also offers new structural convergences and alliances with ethnographic, postcolonial, and science studies approaches. In his contribution, Luisetti focuses on the key arguments of the book, while framing them in the context of contemporary debates on naturalism, post-humanism, animism, and the politics of life, bringing Esposito into conversation with such thinkers as Bruno Latour, Félix Guattari, Eduardo Viveiros de Castro, and Philippe Descola. The main thesis of *Persons and Things*, which builds on the analysis in *Third Person,* is that the long-term juridical-theological logic of the Western episteme, codified by Roman law and Christian theology, is based on separation between things and persons, leaving the body as an unthought, ambiguous, and excluded dimension of human experience, thus letting politics function silently on the bodies of its subjects. Esposito's book, Luisetti points out, "highlights the procedures of purification carried out by the axiological opposition of persons and things, showing how the intermediary of the body—in its manifold occurrences: flesh, animal body, political body, and biopolitical populations—presides over a vertiginous multiplication of splittings and hierarchizations." As such, the body, whether excluded or subordinated, has been placed in an ambiguous middle ground and has guaranteed the translation of persons into things. By denouncing the dualism of Western thought, Esposito suggests that the body may become the decisive element in challenging and rethinking the thing/person apparatus, and in resisting the negative effects of modern biopolitics. For this purpose, Esposito explores some "heretics" of Western philosophy, who have presented alternatives to the dualistic tradition, including Baruch Spinoza, Vico, Friedrich Nietzsche, Henri Bergson, Edmund Husserl, and Maurice Merleau-Ponty, as well as more recent interventions by Gilbert Simondon and Latour. Furthermore, Esposito explicitly attempts to recover nonmodern relations with things, as in ancient Roman law and Maori animistic rituals, thematizing the "archaic and postmodern encounter of persons that are not persons anymore with things that are not things anymore," where the body, as Luisetti puts it, comes to guarantee "the spatial condition of possibility for a new alliance between things and persons, nature and history, science and politics."

In the first essay of the second part of the book, Federico Fridman explores Esposito's category of the impolitical and applies it to a reading of Jorge Luis Borges's literary works. After introducing the reader to Esposito's notion of the impolitical, the essay focuses on an analysis of

Borges's short stories and essays. Fridman shows how, beneath the deepening shadow of nationalist and totalitarian regimes, Borges intensifies and radicalizes politics and its limits in ways similar to Esposito's treatment of the impolitical. He argues that Borges and Esposito share a repudiation of the idolatry of the state and the social that subject individuals to the domination of the political. By bringing Borges and Esposito into dialogue, Fridman contends that for both authors there is an emptiness that dwells within the political, a ground zero, an absolute negativity that defines contemporary politics and all forms of political representation, and so for both authors, politics, as traditionally conceived, remains unable to provide an alternative to the existing regimes of domination.

The concepts of community and immunity are prominent in Mark F. N. Franke's essay that assesses the effects of Esposito's philosophy on International Relations (IR) theory. This contribution challenges the very existence of IR as a discipline by putting in question its basic categories such as sovereign community, international society, law, and politics, as well as humanity and personhood. Franke shows how Esposito's critical reading of Thomas Hobbes's and Immanuel Kant's writings illuminates the limiting and de-politicizing character of the founding premises of IR. First and foremost, IR relies on the assumption of community as a positive political fact, and Esposito's view of community as negativity, as a site of dynamic political struggle rather than preexisting political agency, challenges this condition of international politics. Community is not a given entity that may enter into a relationship with another such entity, but is a project that resists closure. Furthermore, Franke argues, "the positive and contained idea of community, on which IR relies, is thinkable and presentable only insofar as the supposed inside takes in the threats of the supposed outside," which is the mechanism that Esposito describes as immunitary. Last, while IR grounds itself on the universalized premise of human life, Esposito reminds us that the idea of the human, upon which law and specifically rights law relies, does not pertain to the merely living human being, but to the subject of law—the person. In sum, Franke contends, Esposito's philosophy urges us to recognize that we cannot think politics, especially on the international scale, from within such traditional limits as the state, law, community, and the person, but must rather begin theorizing by accepting forms of life that are fundamentally relational and impersonal, making IR theory practically impossible.

Geoffrey Whitehall's essay examines Esposito's theories of immunity and community from the perspective of intercommunity relationship. His insight into the operation of the overlapping mechanisms of the *munus* is centered on a reading of Wajdi Mouawad's *Scorched*—a play that takes place during the Lebanese Civil War and raises the question of how we may live and thrive, rather than merely survive, after trauma, hurt, anger, fear, and loss. Importantly, Whitehall notes, the all-consuming, cannibalistic clash of different communities in the play speaks not only about the particular historical incidents of war, but about the very conditions of modern geopolitical life. Grounded in Esposito's affirmation that immunity and community are endemic, Whitehall's essay allows *Scorched* to stage a different drama of international conflict and security: the well-rehearsed geopolitical imaginary of warring communities is replaced with *warring immunities*. It is not the tragic past or essential differences that condemn communities to clash with each other; they are condemned because they are organized around immunity mechanisms that simultaneously enable and threaten their existence. Whitehall argues that Mouawad's play, in addition to critique, outlines an opening that catastrophe, produced by overlapping warring immunities, presents: the possibility of life's transformation in dramatization, counter-actualization, and becoming political through an opening to the indeterminate temporality of an event. While the latter betrays the immunological sense of causal time that gives us past, present, and future, dramatization enables us to restage the relationships that are involved in having produced the actual, and in this way opening to counter-actualization. Whitehall concludes that "contrary to the promise of Esposito's immunity/community logic, affirmative biopolitics is insufficient to deal with the pain, trauma, anger, and hatred seen in *Scorched*." What is needed is not biological but political rebirth, which Mouawad offers us in the form of his play, allowing us to live through catastrophe.

While Whitehall identifies a performative insufficiency in affirmative biopolitics' capacity to mitigate intercommunity clash, Patrick Hanafin examines its power to transform the immunitary function of the law for the benefit of community. In his analysis, Hanafin focuses on the Italian 2004 Assisted Reproduction Act as an epitome of a negative biopolitics that strives to order and normalize the lives of citizens. He argues that this law reveals a "bio-theopolitics," which both governs and excludes, and thus grounds an understanding of community as "immunity against intruders and ultimately against death, a social compact built on the desire to survive." In the name of the abstract sanctity of life, it disempowers and reduces the freedom of concrete living citizens. However, Hanafin contends, the public challenges to the law that have emerged over the years are indicative of effective counter-politics. The resistant, affirmative biopolitics of living citizens who were denied reproductive choice allows us to speak about a politics *of* life rather than merely *over* life. Citizens' contestation of the law is an example of the "vitalization of the norm": through an ongoing challenge to normalization, they actively engage legal discourse in order to subvert accepted models of subjectivity, community, identity, law, and politics. These citizens fill the abstract norm with concrete vitality, demonstrating that the biopolitical imperative to control and manage life is not one-sided and can be resisted by micro-political acts. Ultimately, Hanafin points out, these acts demonstrate the possibility of an affirmative biopolitics, "a politics which does not valorize an abstract ideologically rigid notion of Life which restricts individual lives, but performs a politics of life which is driven by actions of individual living beings acting in relation with one another." These resistances enact a form of "biocommunity"—a model of community that is based on openness, contestation, and difference. They also provide us with a glimpse of a "creative jurisprudence" where law emerges from the mobilization of self-styling subjectivities.

Notes

1. Among academic journals, see, for example, *diacritics* 36, no. 2 (2006) and 39, no. 2 (2009); *CR: The New Centennial Review* 10, no. 2 (2010); *Minnesota Review* 75, no. 1 (2010); *Law, Culture and the Humanities* 8, no. 1 (2012); *Angelaki* 16, no. 3 (2011) and 18, no. 3 (2013). For recent English publications on contemporary Italian philosophy as well as Esposito's work, see, for instance, Greg Bird and Jonathan Short, eds. *Community, Immunity and the Proper: Roberto Esposito* (New York: Routledge, 2015); Antonio Calcagno, *Contemporary Italian Political Philosophy* (Albany, NY: SUNY Press, 2015); Peter Langford, *Roberto Esposito: Law, Community and the Political* (New York: Routledge, 2015); Alexej Ulbricht, *Multicultural Immunisation: Liberalism and Esposito* (Edinburgh: Edinburgh University Press, 2014); Lorenzo Chiesa and Alberto Toscano, eds. *The Italian Difference: Between Nihilism and Biopolitics* (Melbourne: RE Press, 2009); Silvia Benso and Brian Schroeder, eds. *Contemporary Italian Philosophy: Crossing the Borders of Ethics, Politics, and Religion* (Albany, NY: SUNY Press, 2007). Among notable conferences are "Immunity and Modernity: Picturing Threat and Protection," University of Leuven (2015); "Remembering the Impossible Tomorrow: Italian Political Thought and the Recent Crisis in Capitalism," The British Society for Phenomenology 2013 Annual Conference, University of Oxford; "Commonalities:

Theorizing the Common in Contemporary Italian Thought," Cornell University (2010); "The Impersonal and the Impolitical: The Work of Roberto Esposito," University of California, Irvine (2010).

2. *Vico e Rousseau e il moderno Stato borghese* (Bari: De Donato, 1976); *La politica e la storia. Machiavelli e Vico* (Napoli: Liguori, 1980); *Divenire della ragione moderna. Cartesio, Spinoza, Vico*, co-authored with Biagio De Giovanni and Giuseppe Zarone (Napoli: Liguori, 1981); *Ordine e conflitto. Machiavelli e la letteratura politica del Rinascimento italiano* (Napoli: Liguori, 1984).

3. *Nove pensieri sulla politica* (Bologna: Il Mulino, 1993); *L'origine della politica. Hannah Arendt o Simone Weil?* (Roma: Donzelli, 1996).

4. Massimo Cacciari, "L'impolitico nietzscheano" (1978), published in English as "Nietzsche and the Unpolitical," in *The Unpolitical: On the Radical Critique of Political Reason*, ed. Alessandro Carrera (New York: Fordham University Press, 2009), 92–103.

5. Roberto Esposito, "Community, Immunity, Biopolitics," *Angelaki: Journal of Theoretical Humanities* 18, no. 3 (2013): 85.

Part I

READING ESPOSITO

I

The Presence and Absence of Origin

Roberto Esposito's Early Interpretation of Giambattista Vico

Alexander U. Bertland

Giambattista Vico (1668–1744) presents a striking account of the origin of thought. In his *New Science*, he paints a dramatic picture of a state of nature in which giants, descended from the children who abandoned Noah, wander in an animal-like fashion around the forest.[1] They suddenly hear a giant thunder strike, and they cannot instinctively recognize its source (*SN*, 377). This noise, which they imagined as a commanding voice from the sky, awakened both their powerful imaginations and an intense moral compulsion (*SN*, 504). Hence, a profound superstition caused the first people to settle and found the first communities. This fanciful account of the origin of society puts Vico in sharp contrast with the more traditional social contract theories of Thomas Hobbes and John Locke.

Roberto Esposito emphasizes in his later writings the importance of the gap Vico delineates between the bestial and the social. In *Immunitas: The Protection and Negation of Life*, Esposito refers to the work of Gennaro Carillo, which emphasizes the Vichean divide between the undifferentiated masses of giants (*giganti*) and the law that allows humans to take on a political form (*IM*, 43–44). Vico separates these two realms to explain that the law's function is to maintain a rational order that exists apart from the feral state. At the same time, the separation between the realms reveals a constant threat that the civil world could return to the wild state of roaming masses of giants. In his work *Living Thought: The Origin and Actuality of Italian Philosophy*, Esposito explains that, for Vico, the course of history must always fight against the nonhistorical wilderness in order to maintain its order, yet it cannot eliminate the nonhistorical since that provides the course of history with its energy. Esposito writes, "Nothing is more deadly, for Vico, than the typically modern idea that we can never sever the knot that binds history to its nonhistorical beginning, unraveling it through a process that fully temporalizes life" (*LT*, 27). In Esposito's later work, this one moment becomes the main source from which Esposito draws insight from Vico.

At the beginning of his career, Esposito had a much broader interest in Vico's *New Science*. Two of his earliest works centered directly on Vico: *Vico e Rousseau e il moderno Stato borghese* published in 1976 and *La politica e la storia: Machiavelli e Vico* published in 1980.[2] In these texts, Esposito discusses the way Vico reconstitutes the connection between the metaphysical and historical, but they do so in a way that goes beyond the feral/social distinction. The works broadly analyze how Vico conceives a metaphysical intertwining of the historical with the political realm. Vico, throughout the *New Science* and beyond, plays with the intersections and confrontations of institutions or, as Esposito would later call it, the "immanentization of antagonism"—in order to reveal the metaphysical truth that underlies these interactions (*LT*, 24). Within a variety of oppositions in history, Vico reveals a profound sense of order that is lost on Machiavelli and Rousseau.

This chapter will examine these early writings to illuminate some key elements of the Vichean relationship between the metaphysical and the historical that Esposito identifies. Unfortunately, the main reason for returning to these early works of Esposito cannot be the content of the analysis. Esposito does not repudiate these early works, and he always maintains that Vico is important. At the same time, however, he stops drawing on specific ideas from Vico's account of history. At the end of the chapter, I will speculate as to why this might be the case.

More specifically, this chapter, explores how Esposito's approach in these early texts served as an initial testing ground for his later philosophical approach. In his later writings, when Esposito investigates a social institution, he recenters one's orientation toward it, shifting what is essential and what is accidental. He recontextualizes what is foundational and what is arbitrary. He does not simply show how established moral norms are actually in flux; he reveals how previously hidden necessities are guiding this historical process. For example, in his recent work, *Due: La macchina della teologia politica e il posto del pensiero*, Esposito shows that the apparent division in liberal democracy between religion and politics does not really hold, and that other factors serve as boundaries that guide political understanding. This play of historical institutions and hidden necessity can be seen throughout Esposito's later writings.

Esposito found in Vico's *New Science* a philosopher who made the same movement that Esposito himself would later make. In Esposito's interpretation, Vico used the evidence of historical analysis to reshape the boundaries between the essential and accidental in political institutions. The *New Science* appears outlandish and bizarre but this is because, as Esposito notes, Vico recognized the need to adapt his language to reflect the mentality of a different historical era (*LT*, 74). Nevertheless, the *New Science* represents a legitimate attempt to research empirically the development of human political institutions at an epistemological level and reveal hidden structures within them. Vico discovered a gap between prephilosophical poetic wisdom and a post-philosophical rational conception of the world. The original conflict between these two forms of thought established tensions that governed social development and the growth of political institutions. Esposito finds these tensions in Vico and he explicates them in comparison with Rousseau and Machiavelli.

An essential point of Esposito's analysis is that Vico's examination of the tension between these forms of thought never reduces history to historicist arbitrariness. On the contrary, Vico always insists that within the course of a tumultuous history it is possible to find a coherent and rational order that supports the civil world.[3] This becomes both the philosophical and methodological link between Vico and Esposito. Esposito later loses interest, apparently, in Vico's Baroque account of the development of Roman law. Nevertheless, Esposito shares Vico's desire to locate determinate structures in the growth of political institutions. It is worth dwelling on Esposito's early ideas about Vico's approach to philosophy to consider how this may have shaped his later approach.

Methodologically, both thinkers approach philosophical writing using a similar set of tools. Their texts are quite different because Esposito draws on a wide range of twenty-first-century research while Vico had to rely on the Baroque and ancient texts that were available to him. Nevertheless, because both authors investigate conflicting and intertwining political institutions and structures of thought, they reject a systematic approach to philosophy that relies on a linear derivation of conclusions from premises. Instead, their works weave and circle across many topics in order to draw together disassociated points while separating ideas held in common. Vico presents a series of axioms (*degnità*) that provide justifications for his *New Science*. Yet, these axioms do not serve as first principles as they might for an author like Spinoza. Instead, they are meant to course through the entire science (*SN*, 119).[4] Further, Vico is fond of using oxymorons to identify the fundamental ideas of his science, most notably "poetic wisdom" and "ideal eternal history."[5] From a traditional Enlightenment perspective, the connection of these terms does not make sense, but Vico uses them as foundational points in his science, connecting what is normally unconnected. Perhaps most importantly, Vico and Esposito share a passion for etymology. Esposito's analysis of community and immunity depends on his etymological analysis of their roots in the term *munus* (*IM*, 5–6, 22–23). Vico's most famous idea, that the true is the made, depends on his etymology of *verum* and *factum*.[6] His *New Science* is filled with etymological analyses of words such as *padre* and *urbs*.[7] The imaginative exploration of word origins by these two authors should not be confused for a lack of rigor. Both authors take great care to provide consistency within their projects. This is harder to see in Vico because of his dependence on ancient sources and the separation between the eighteenth century and the current day. Nevertheless, Vico certainly worked to present a coherent science. The fact that these authors used these similar tools reveals their shared commitment to the idea that the metaphysical may be uncovered in temporal institutions.

Having identified this underlying similarity, it is possible to explore the content of Esposito's early interpretation of Vico's *New Science*. Esposito has a distinctive way of placing Vico's ideas in a historical context. Many Vico scholars and Vico himself present his work as being in opposition to the greater scientific project of the Enlightenment and specifically to Descartes. The extent to which Vico should be understood as anti-Enlightenment has recently become a more open question.[8] Regardless, Esposito side-steps that discussion by putting Vico in conversation with Rousseau and Machiavelli. These authors share many of Vico's themes. In particular, in the *Discourse on the Origin and Basis of Inequality Among Men*, Rousseau presents an account of history that is very similar to Vico's. In fact, one might suspect, despite the lack of evidence, that Rousseau encountered the work of Vico while in Venice, the one city where Vico's work had a major influence.[9] Machiavelli shares Vico's desire to locate his theory directly in human political institutions as epitomized in the history of Rome. While other scholars have discussed the connections between Vico, Machiavelli, and Rousseau, Esposito uses this comparison to focus primarily on the question of how historical investigation ought to relate to philosophy.

Like Vico, Rousseau and Machiavelli accept that a philosopher needs to work from the evidence of history. Further, all three thinkers discuss possible courses of history albeit with different levels of metaphysical commitment.[10] They share a concern about the grand structure of the course of historical development. Esposito could have contrasted the ways they charted different paths of history, but he did not. Instead, Esposito examines the commitments that are entailed by their different approaches to history itself. All three authors, as thinkers who explore history philosophically, are committed to the notion that within particular historical events, universal truths can be located. They differ in how they portray the relationship of the universal to the particular within the course of history. This becomes the focal point of Esposito's investigations.

Esposito makes this question explicit in a passage of *La politica e la storia* through a critique of the hermeneutic reading of Vico. Hermeneutics has valued Vico's discovery of a *sensus communis* that is established by nonreflective but linguistic thought and that serves as a context for communication.[11] Hans-Georg Gadamer emphasizes this aspect of Vico in *Truth and Method*.[12] According to Esposito, Gadamer's approach is flawed because it fails to explain the significance of Vico's science of history. In Gadamer's reading, Vico draws a sharp divide between practical wisdom and theoretical wisdom and between the abstract and the concrete (*MV*, 170). Gadamer does not attempt to explain or bridge this gap. Practical wisdom can look to the abstract to understand its own limit and finitude, but it cannot enter the theoretical. Theoretical wisdom is entirely blocked from the practical.[13] Esposito writes that in Gadamer's interpretation, "the expression *science of history* is a true oxymoron, a contradiction in terms, a theoretical paradox: at least if the term 'science' remains charged with its traditional, and frankly undeniable, rational character . . ." (*MV*, 171). Esposito is unsatisfied with Gadamer's approach, and he tries to explain Vico's true value, which may be found in his science of history that unites theoretical and practical wisdom.

Esposito examines the way in which Vico connects particular historical instances with the universal pattern of history, which Vico calls the ideal eternal history. The traditional reading of Vico is that the ideal eternal history is an abstraction into which all particulars may be placed.[14]

Esposito warns that this view makes the assumption that particulars are static and that their dynamic relation is explained through the universal pattern. Esposito emphasizes that particulars are themselves temporal and therefore dynamic. It is the historical flow of the particulars that give rise to the universal pattern of history. Instead of starting with the universal pattern and placing particular events into it, the universal grows over time out of the particular. To understand the universal pattern and to develop theoretical wisdom, therefore, it is necessary to begin from practical wisdom and the historical existence of the particular. This is the basis for Esposito's most fundamental recentering of Vico's thought.

Esposito locates this new relationship in Axiom XV of the *New Science*, "The inseparable properties of institutions must be due to the modification or guise with which they are born. By these properties we may therefore verify that the nature or birth was thus and not otherwise" (*SN*, 148; *MV*, 274). An institution's particular origin reveals what is necessary about it. One may not separate an object from its temporal stream and still unveil its essential nature. One must grasp the temporality of the object to recognize its universality. Vico has a science, but it is one in which the metaphysical must be revealed directly within the course of history.

While Esposito does not develop this particular point, Vico's notion of divine providence provides a way of bringing the theoretical and the practical together within the science. Vico insists that the motion of bodies is the result of the free choice of the will (*SN*, 340). Yet divine providence orchestrates these choices such that they sometimes advance the course of history, despite the intentions and objectives of those making the choices. To understand how divine providence operates, one should not just look at what individuals thought they were trying to achieve. Vico stresses that finding the metaphysical truth of divine providence requires one to look carefully at the essential aspects of institutions at their origins. Vico writes, "Compare the institutions with one another and observe the order by which those are now born in their proper times and places which ought now to be born, and others deferred for birth in theirs . . . Such proofs eternal wisdom provides" (*SN*, 344). By looking at the particulars in history—specifically the origin of an institution and the way it unfolds—it is possible to reveal metaphysical truth.

Commentators on the *New Science* tend to focus on Book II in which Vico outlines poetic wisdom as a unique form of consciousness.[15] Esposito focuses his reading of the *New Science* on Book IV in which Vico presents different human institutions as they existed in each of the three ages (*VR*, 97; *MV*, 274). There Vico combines his account of poetic wisdom together with the particular account of Homer and Greek history that he presents in Book III (*SN*, 915). Esposito is interested in the way Book IV shows how the particular origin of an institution leads necessarily to the permutations that follow. Esposito writes:

> Book IV, the last and densest section of the entire work . . . furnishes a *profound* canvas, adapted to reconstruct not only the connection, for Vico, hidden between universal and particular, but . . . the spontaneous mode with which the universal is born and grows within the particular. This frees the latent element—*facilitas*—that makes the particular the natural root of the universal and the universal the rational product of the particular, flowing (and consuming itself) along a line that goes from ignorance to awareness." (*MV*, 274)

In Esposito's reading, Book IV does not just show particular institutions categorized into different ages of a universal history. Instead it shows how particular moments of origin shape the universal categories across the flow of history.

This reading of Vico surprises those familiar with Book IV. Vico's ideal eternal history is a cyclical series of three ages through which all nations pass: gods, heroes, and humans. In Book IV, Vico outlines how certain institutions, such as government, language, and jurisprudence, change through the three ages; each institution is presented in three distinct forms in each different age. This book appears the most categorical of the five because it seems to delineate boundaries that separate the ages. According to Esposito, reading the three ages as distinct categories obscures the organic spirit of Vico's system. Vico's goal is not to conceptually draw boundaries between ages but to understand each particular institution as it exists in history through its development across the ages. Vico's point is not to explain the difference in ages but instead to understand each institution as the universal history plays through the particular social structure. This, according to Esposito, is a unique attempt to investigate history through metaphysics.

This point is supported by the fact that Vico rarely asserts strong divisions between the three ages. When he presents the ideal eternal history in the axioms, he does not always articulate three clearly delineated ages. Instead he presents it in alternative ways that sometimes have five or six different stages (*SN*, 239–43). Further, in Book IV he dedicates a section to the point that in the transition from an old age to a new one, the old governmental administration remains for a period (*SN*, 1004–06). This section may serve a number of functions for Vico, including an attack on the idea that mixed governments are politically viable. Nevertheless, it clearly shows that Vico was not concerned with defining clear borders between ages. This supports Esposito's interpretation that the ideal eternal history should not be understood as a set of abstract temporal concepts. Instead, Vico starts from examining the historical origin of particular institutions and tracing their organic development rather than articulating the ages as definite categories. Thus, Vico never fully separates theoretical wisdom from the practical investigation of the development of historical institutions.

Esposito contends that Vico's connection of history and metaphysics allows him to find and endorse courses of political action like those of Rousseau and Machiavelli. This becomes clear in the specific comparisons Esposito makes between Vico and each of the two other authors.

There are key differences between Esposito's works on Rousseau and on Machiavelli. The earlier work on Rousseau reveals a more overt interest in Marxism (*VR*, 10–12). It also emphasizes the fact that for Vico the driving force of political systems, from their very origin, is violence.[16] In *Living Thought*, Esposito returns to the significant role of violence in Vico's thought (*LT*, 79–81). Nevertheless, both works have temporality as their central concern.

Esposito begins his work on Rousseau by contrasting the way Rousseau and Vico understand the impediments to self-knowledge. Both authors write an autobiography, but the problems they claim to face are different. Rousseau struggles in the *Confessions* to express his true passionate self as it exists outside of the political world.[17] In his *Autobiography*, Vico is unconcerned with exposing his hidden personality; instead, he connects the pattern of his life to the ideal eternal history.[18] Rousseau seeks the truth at a level beneath the surface while Vico connects the surface of his life to metaphysical truth. One can find many contextual reasons to explain the difference between the two authors, including the fact that in France there was a long tradition, reinforced by Jansenism, of seeing truth outside of human action, while the intellectuals in Naples rejected this tradition.[19] Instead of focusing on cultural influences, Esposito locates the difference in the authors' diverse conceptions of temporality.

In Rousseau's writings, Esposito finds the impediment to self-knowledge as the density of the flow of time itself (*VR*, 23–24). Rousseau wants to reveal an eternal self that exists outside the accidental changes of time. He writes the *Confessions* not to see himself in history but to find the hidden threads within himself that underlie all his authentic actions. The superficiality of the present and the flow of the world of becoming prevent him from seeing what he is.

Conversely, Vico, unconcerned with revealing a hidden self, suggests that the vividness of the present prevents one from seeing the self as connected to its origin and within a temporal stream. The immediacy of the present causes one to make abstract assertions about the self that become the core of a false self-knowledge (*VR*, 22–23). The present clouds the metaphysical, but this is because the metaphysical can only be revealed as it exists in the course of history and the immediacy of the present hides the fullness of this temporal stream. The only way to reveal the metaphysical is to place the immediate present in its historical context. This historical contextualization allows one to distinguish what is necessary from what is accidental in the immediacy of the present, exposing its metaphysical dimension. This fits with Esposito's account of Vico's science above.

Esposito then reverses the traditional reading of the ethical positions of the two authors. Vico does not present a clear vision of a model society, and some interpreters are frustrated because he does not express clear moral imperatives.[20] Rousseau, on the other hand, does not hesitate to make moral prescriptions, albeit often inconsistent ones. Despite this, Esposito argues that Vico's understanding of time and metaphysics puts him in the better position to give guidance on practical wisdom.

Because Rousseau has divided the self from the course of time, it is impossible for him to explain how a political community, which is in the course of time, can have a substantial positive effect on the self. The individual, on Rousseau's view, strives to be outside of time in order to have an authentic sense of self. Esposito writes that for Rousseau, "Somehow consciousness (*coscienza*) is not what is at the origin, but the origin is in consciousness" (*VR*, 140). Actual differences between individuals lie within their consciousnesses as they exist apart from the flow of time. The temporal origin of consciousness, as developed in society, only produces inauthentic differences. The true origin of identity is not temporal but lies fixed within conscience itself. As a result, the passage of time moves the individual away from that origin and threatens to separate the self from the origin. This places Rousseau in a classical tradition that fears the passage of time and its threat to the self; Esposito writes, "In regards to time, Rousseau fears the future" (*VR*, 142). The only response to this threat is for the individual to bring the future back into the present. The individual must come to recognize that the fear of losing oneself in the future is, in fact, a present concern. By locating this concern in the present rather than the future, the individual returns to its origin in conscience that is linked to the present but outside of time. In this way, the individual can be preserved and the future may be stopped (*VR*, 145–46).

For Rousseau, then, the realm of the political cannot enter into the origin of consciousness. The political is always in the temporal because it always involves change. The general will, as organized by the social contract, is made up of a variety of particular wills and is supposed to assure their freedom over time. It cannot escape the flow of time. Yet, the social contract cannot project into the future without impeding the freedom of the particular wills that are outside of time. Thus, the general will is blocked from functioning effectively. The sovereign can pass laws, but over time those laws will always be questioned by particular wills. Esposito writes, "The latent force of particular wills creates and imposes the continual actualization of the law that, emanating from the general will, would naturally have the right to a lasting force, but one finds instead that they continually put the law into discussion" (*VR*, 153). As soon as a law is passed, circumstances change the needs of the particular wills and the law will be questioned. The sovereign cannot enforce lasting equality within a political society, which is what Rousseau was hoping the social contract would achieve. The only remedy that Rousseau offers for this is to separate the private from the public in order to allow for equality in the public realm. This, however, is too great of a sacrifice since the loss of equality in the private realm would stray even further from the desired effect (*VR*, 154). Since the laws as they exist in time cannot organize or guarantee freedom of the wills outside of time, the laws of the sovereign cannot serve to benefit substantially one's actual private will.

According to Esposito, any attempt to resolve this problem through the deployment of ideology cannot succeed. Ideology must originate as a temporal manifestation from particular wills that are always unequal as they manifest in time (*VR*, 156). There would need to be an original equality to produce an ideology of equal rights, but such an original state is not possible (*VR*, 160). Ultimately private interests, and specifically private property, will be placed before the general interest. As a result, given Rousseau's characterization of time, true democracy cannot protect a sense of equality (*VR*, 167).

Esposito asserts that Vico does not face these problems. For Vico, the self does not have an atemporal aspect. In fact, the metaphysical understanding of the self comes from seeing how the self exists at its temporal origin and across the flow of time. Esposito characterizes Vico's view in this way, "The past continually contradicts the present, perhaps gradually conquering it, threatening to drag it behind: but this in itself demonstrates the presence of the past (certainly not all the past) in the present, the relative continuity that hooks them together, the substantial homogeneity that constitutes their connection" (*VR*, 31). Because of this connection between the present and past, the temporal and metaphysical, there is a way for practical wisdom to work within the flow of time that positively influences the individual.

Esposito clarifies the aforementioned point by presenting the contrasting views of Rousseau and Vico on utility, honesty, and ideology. For Rousseau, utility, at least when it is manifest in a societal structure, obscures honesty because utility must be based in the temporal. Drawing on Vico's fourth inaugural oration, Esposito claims that for Vico utility is never in discord with the honorable (*VR*, 45). There is a complex relationship between utility and honesty because they exist at two different levels: utility is corporal while honesty is abstract. Nevertheless, for honesty to have value, it must ground itself back into the temporal world of the corporal. Esposito writes, "It is true that while utility does not constitute a point of support for honesty nor its internal motivator, and yet where the parabola of honesty concludes, where its measurable ark stops, utility inevitably reemerges."[21] On Vico's view, as society advances, it becomes more of a meritocracy. In earlier ages, leaders could rule through pomp and circumstance. Over time, the people come to see through this. As a result, the leaders need to become useful for the lives of the people (*VR*, 47).

The practical implications do not stop there. Honesty requires the ability to think abstractly because it requires one to be able to reflect on what one is saying. This abstraction was not possible for early thinkers who used poetic wisdom that is dominated by mythic images. As humanity transitions into the third age, it becomes necessary to raise people's ability to think abstractly so they can be honest (*VR*, 49). Leaders of the age of humans have a moral obligation to raise the intellect of the people to the level of honesty. Somewhat counterintuitively, this creates a positive need for ideology. On one level, ideology is based in verisimilitude that is an intermediate step between falsehood and truth (*VR*, 60). It serves as a ground for the development of civil prudence and social cohesion.[22] Rousseau could not endorse the use of ideology for the masses because he could not deny the supreme importance of preservation of the integrity of individual selves. Even at the end of the *Social Contract*, where Rousseau calls for the deployment of a civil religion, he does not want that religion to interfere with freedom and self-knowledge.[23] For Vico, who sees the present as grounded in the past and can see institutions in a greater social perspective, ideology can push farther as long as it carries individuals toward a greater sense of civil responsibility.

According to Esposito, Vico's system can ultimately produce a system of justice. Rousseau sought *iustitia aequatrix*, an arithmetical understanding of equality that strives to preserve for everyone an equal value. This is based on a vision of equality grounded in a particular slice of time. Vico, on the other hand, sought *iustitia rectrix*, a geometrical sense of justice that strives to distribute goods proportionally (*VR*, 103–05). Both theories call for the redistribution of wealth. Vico's theory grounds itself in the course of history and so develops slowly by initially maintaining the existing dominion of the upper class and slowly spreading the wealth more thoroughly across the newly defined public sphere (*VR*, 105). The historical forms of power are maintained but come to be relegitimized through the development of a new relation between the leaders and the people. The transition between the feudal age of heroes and the reflective age of humans amounts to a reconstruction of the relationship between the public and private sphere (*VR*, 115–16). In the age of heroes, rulers held the public interest privately in the sense that their personal interests drove the actions of the public. In the age of humans, the private sphere grows as the old heroic and feudal system decays. As this happens, a monarchy develops that does not impede the growth of the private but is charged with the task of directing it in order to facilitate the fulfillment of the public interest.[24]

Thus, for Vico, it is possible to achieve the equality and justice that Rousseau sought. If bourgeois leaders take responsibility and educate their people, they can create a just society. If the leaders take seriously their duty to fulfill the requirements of utility and develop social cohesion through the development of ideology, then society can flourish (*VR*, 128). Because the metaphysical is in the particular, a proper course is inevitably charted and it is up to the free will of the leaders to follow it. Hence, there is a moral imperative for Vico. In Rousseau, conversely, there always exists a gap between revealing the self and effecting the political change that makes progress impossible.

Esposito has a unique way of understanding the conversation between Vico and Rousseau. Because he locates a difference in how the authors relate the self to the course of history, Vico, rather than Rousseau, becomes beneficial for developing practical wisdom. Esposito develops an even more exceptional way of contrasting Vico with Machiavelli. Esposito brackets any discussion of reason of state. He also disregards Vico's two actual criticisms of Machiavelli in the *New Science*: that Machiavelli gave too much credit to the survival of the Roman republic on the plebs, and that Machiavelli has too mechanistic an approach to history (*SN*, 1003, 1109). Esposito is interested in the way Machiavelli connects knowledge to power and the way he develops a new science of ideology, but his main focus is once again the different views of temporality held by the two authors (*MV*, 85).

When examining Machiavelli's political philosophy, the challenging question is whether he actually identifies universal moral norms that should limit the power of the prince. Esposito locates these necessary norms in an unexpected place. In the *Discourses*, Machiavelli explains how the development of an ideology can maintain a principality. The effectiveness of the ideology depends on the fact that the masses experience ideology in the context of the historical order while the person deploying the ideology does so atemporally from a scientific distance (*MV*, 92–95). The limit of a government's power is established by the limit of the government's ability to present an ideology. Esposito suggests that Machiavelli takes this idea to its logical conclusion in *The Prince*. There, Machiavelli identifies the limit to the effectiveness of ideology as death. Machiavelli dwells on the downfall of Cesare Borgia, who would have been in a position to unify Italy had he not been sick when Pope Alexander VI died (*MV*, 101–04). For Esposito, this is the defining example for what he calls Machiavelli's philosophy of crisis. Death continually threatens to strike and the prince's success depends on being able to spot and circumvent crises before they can cause too much damage and undermine the ideology that holds a community together.

Machiavelli's advice to the prince is similar to the advice Rousseau gives to one seeking self-knowledge. For Rousseau, an individual's true self always rests outside the temporal flow. For Machiavelli, the prince must rise out of the flow of history. The threat of crisis and death does not allow for any sort of mediation. The only response the prince has is to stop the flow of time or to adapt to it by taking a perspective that is unaffected by time. Esposito writes, "For being powerful, the politician *must*—behold his momentous destiny—put himself outside of history. His state has a history (even if it is obsolete and unproductive), *it is* history; the politician is he who founds it, crosses it, but does not live it: he is outside of historical representation, historically unrepresentable as an object of history."[25] The threat of crisis makes time the enemy of the prince; the prince must escape time in order to adapt to it, change it, and benefit from it.[26]

This is seen in numerous examples of Machiavelli's practical advice to the prince. He suggests that when a new state is acquired, the prince must radically break that new territory from its previous historical path. There is no possibility of mediating old traditions and new institutions, so the chain of time must be immediately broken, thus breaking the flow of history (*MV*, 117–18). Machiavelli praises the Romans for their practical ability to anticipate problems in the future. This requires stepping out of the rhythm of time to recognize the need to adapt to the future (*MV*, 125). Machiavelli strongly emphasizes the need for the prince to always make and enact decisions that are swift and decisive. The prince should not passively adapt to changing circumstances but should instead escape the flow of time, recognize the need for decisiveness, and then abruptly alter history to maintain the security of the principality (*MV*, 129).

The philosophy of crisis is at the core of Machiavelli's claim that a mixed system of government is best. Since a crisis may come from any aspect of a state, the prince ought to maintain a government that contains different aspects of society—democracy, aristocracy, and monarchy. The leadership may then use these various perspectives to adopt an atemporal view of changing conditions within the state. The variety of available perspectives prevents leaders from being constrained by the flow of time and allows them to rise above it by shifting from one view to another.

Notably, Machiavelli's philosophy of crisis alters the traditional notion of fortune (*fortuna*). On the ancient view, fortune is thought of as an objective force that negatively or positively influences an individual. For the ancients, no one can be expected to anticipate or avoid the strokes of fortune. Sometimes these strokes of fortune can bring great benefit and other times tragic downfalls. On Machiavelli's view, acts of fortune are nothing more than chance occurrences and there is no inherent meaning or value in them. It is a leader's responsibility to prepare for and respond to these chance occurrences. Machiavelli is insistent that Rome did not grow its territory on the basis of fortune. There were fortunate incidents, but the republic grew because it had the foresight to take advantage of those opportunities. Conversely, when a leader cannot take advantage of fortune or is thwarted by it, this represents a lack of *virtù* on the leader's part.[27] Fortune in itself cannot have any real meaning or significance.

According to Esposito, Machiavelli's philosophy of crisis must ground itself in a particular understanding of the philosophy of history. In order for random chance to play such a powerful role in the decision-making process of the prince, Machiavelli cannot allow for there to be a greater purpose or pattern to history. Instead, history must be nothing more than a chain of causal events.[28] Dramatic events of history can only serve to undermine stable communities.

Vico attacks Machiavelli's account of temporality because it does not attempt to find a metaphysical significance to history. Esposito writes:

> What Vico argues, every time that he laments the incapacity of Machiavelli in particular to understand the sense, the "essence," the "cause" of Roman history, it is not so much the presence, perhaps inadequate, of a philosophy of history, as much as, literally, its absence: neither Machiavelli, nor Hobbes—even if both make continual reference to historical objects and dynamics—succeed[s] to categorically conceptualize the term "history." (*MV*, 272)

In other words, Machiavelli has failed to think theoretically about history and so he is restricted to the realm of practical wisdom.

At the root of Machiavelli's failure to look at the philosophy of history is a superficial understanding of the way in which a historian should represent the evidence of history. Esposito writes, "in Machiavelli, representation overlaps with time, but in a form that, instead of adapting the categories of representation to the movement of time, definitely subordinates the movement of time to representation" (*MV*, 272). Machiavelli recognizes that the historian must work to represent change over time but not that this change calls for the alteration of the form of representation itself. He thus modifies the evidence of history to fit his way of thinking, thereby flattening out differences in temporal periods. Although Esposito does not refer to it here, Vico would accuse Machiavelli of the conceit of scholars, which holds that all people have always thought as contemporary historians do (*SN*, 127–28). Conversely, according to Esposito, "History, in Vico, breaks down representation, gives it weight, body, and profundity. It gives it another dimension, doubles it, multiplies it" (*MV*, 273). Vico recognizes that in order to understand history one must enter into the theory of representation itself. Vico prioritizes the movement of history over representation and so works to adapt his manner of representation to suit the changes in history. Vico recognizes a need for a philosophy of history that Machiavelli always ignores.

By developing this theory of representation, Vico can trace the entire course of history. This provides for a metaphysical stability to history that Machiavelli's philosophy of crisis lacks. This is what Esposito emphasized when he praised Book IV of the *New Science*. By recognizing the vertical course of history, Vico understands that its metaphysical structure will maintain a necessary flow. In opposition to the philosophy of crisis, Vico thus proposes a philosophy of preservation (*conservazione*) that holds that society develops along a bounded historical path (*MV*, 263).

There is a temptation to understand the ideal eternal history as one of alternating periods of crisis and stability, punctuated by a dramatic collapse. Esposito rejects this interpretation in favor of seeing in Vico a philosophy of preservation. When Vico presents the barbarism of reflection and the ensuing rebirth of society, he praises divine providence for preserving humanity.[29] While society should work to forestall the coming barbarism of reflection, one should recognize that its inevitability has as its primary purpose the preservation of humanity rather than its complete destruction.

Within the course of the ideal eternal history, transitions from one age to the next are not dramatic turning points of crisis or sharp revolutions that leap from one stage to the next. Unlike Machiavelli, Vico does not dwell on the assassinations of the Roman king or Julius Caesar as moments of transformation. For Vico, the order of ideas has to follow the order of institutions (*SN*, 238). When there is a movement between ages, social change occurs before there is a full awareness of the change. This means change will always be slow because it will take time for institutions to evolve and ideas to manifest the change. While there are distinguishing features to the ages, the course of history moves in an organic flow, as discussed earlier in the chapter.

This insight is significant not just for understanding the course of history itself but also for thinking about how historians should approach their subject. Rather than focusing on dramatic innovations, they should explore how the institutions of one age necessarily constitute the institutions of the next. This relates back to the above-mentioned discussion of Book IV of the *New Science*, but Esposito also locates this claim in Vico's *De Constantia*. For example, when the Roman plebs request equal laws, this may only be understood by understanding the specific manner in which the plebs were oppressed in the previous era. The event which marks the birth of the age of humans can only be understood in the context of the previous heroic era. Esposito writes, "This truly expresses the theory of history: *the recognition of the plurality and specificity of discrete historical eras and the awareness that the logic of the determination of their contents is directly dependent on the form of their connection*" (*MV*, 226). Not only do transitions in history need to be lived organically, but historians must understand them organically. The only way that the historian may understand an era is by shedding light across different eras to reveal the connections between them.[30]

Esposito describes the disagreement between Vico and Machiavelli in this way:

> With Machiavelli—in a phase at least at the genesis of bourgeois ideology, and still contesting at length the residue of pre-bourgeois elements—the point of view from which to examine history is that of crisis, of the closing of a cycle, of the radical end of times; for Vico it is that of balanced development, of a flow between limits, of a regulated process; the closure of one phase is *already* the opening of the next, and nothing of the past is lost definitively, is consumed, is burned, is depleted. (*MV*, 296)

This leads to profound practical implications that separate the two thinkers. Vico's leader, in the third age, will have a great deal more flexibility because there is not a constant shadow of crisis. Once the course of history is understood scientifically, leaders will grasp the general flow of the course of events and will be able to use this knowledge to maintain stability for their political states. This will empower them to create, within the boundaries of the temporal flow, institutions that encourage justice. Rather than a Machiavellian leader who takes whatever means necessary to anticipate crisis, the Vichean leader of the third age will privilege freedom through the promulgation of philosophical investigation knowing that as long as they stay within the bounds of the course of history there will be political stability.

Vico rejects the Machiavellian proposal that there can and should be mixed forms of governments. Instead, Vico holds that each type of government should exist at its own time. On one level, since there is not the concern for crisis, Machiavelli's call for having a mixed form of government to anticipate crises is unnecessary. More deeply, because Vico has a richer understanding of the connections between different historical ages, he better comprehends the links and separations between different forms of government and modes of thought. Since it is impossible for a later form of thought to recapture the mentality of an earlier age, there should be no attempt to bring an earlier form of government back into the present (*MV*, 292–93).

Instead, Vico turns to monarchy as the one proper system of power for the third age. To understand this, however, one must see Vico in Esposito's full inversion. When one ordinarily thinks of monarchy—perhaps in the sense of Hobbes or Machiavelli—one thinks of an absolute ruler standing outside of the people and maintaining order. This is based on a loose religious paradigm that suggests that a strong religious leader is necessary for maintaining order. For Vico, in the ages of myth, religion was extremely powerful but this is not because leaders consciously imposed religion to maintain order. The leaders were as cowed by the superstition as anyone. As the need for religion was lifted and replaced by the desire for liberty, everyone comes to desire freedom and justice. In the third age, the leader shares the same mentality with the people (*MV*, 287).

The purpose of government in the third age, therefore, is to maintain a balance of social control and freedom. It is not, in a Machiavellian sense, to maintain order in a paternalistic way. In fact, Vico wants to diffuse a sense of freedom through the people as a way of encouraging equilibrium. Esposito writes, "Vico is not a free-civil libertarian philosopher of the particular, but neither is he a systematic philosopher of the state machine (as Hobbes, to be clear, but not for Rousseau). The two demands are mediated, appeased, resolved: at least to the point in which the always latent dualism explodes with all of its destructive force" (*MV*, 283). Esposito links Vico's political philosophy to the development of a bourgeois mentality. Vico's politics works to serve as a justification for a community in which the private and the public are organically conjoined in which all are working to expand freedom. The growth of liberty occurs with the security of knowing that even when society inevitably implodes, society will begin again.

There is much more to be said about Esposito's claim concerning Vico as a bourgeois thinker. Instead of exploring this claim, this chapter will conclude with some speculation as to why Vico's role in Esposito's later work has diminished. Before this speculation, however, it is important to remember that Esposito does not appear to disown Vico or his earlier writings on Vico. Esposito consistently praises Vico throughout his work. Further, Vico's point, discussed at the opening of this paper, that there is a separation between the feral and political appears to hold an essential role in Esposito's later ideas about law and community. Esposito seemingly has left the door open to return more substantially to Vico.

Nevertheless, Esposito may have encountered a problem that many of Vico's readers face. Vico was dedicated to integrating metaphysics with actual historical evidence. Unfortunately, despite what many think were Vico's best attempts, there were a large number of things Vico got wrong.[31] When working with Vico, one has to decide whether to get into the minutiae of Vico's broken history or leave much of the content of Vico's work behind. Esposito, who himself wants to use solid historical evidence, has seemingly chosen the later. While Esposito does not draw attention to it, there is a major division between his ideas and Vico's in *Immunitas*. Esposito there carefully examines Roman jurisprudence to get at the difference between community and immunity, but he does not refer at all to Vico, whose training was primarily in Roman jurisprudence. Instead, Esposito's starting point is the work of Rudolf von Jhering who holds that, in ancient Rome, law preceded religion rather than religion preceding law. This is fundamentally contrary to Vico's account of Roman jurisprudence since, in the *New Science*, Vico clearly places the religion of Jove before the institution of law (*IM*, 27). On points such as these, it becomes difficult to include Vico in the conversation. Nevertheless, Esposito makes clear that he respects the contribution of Vico to Italian philosophy and his own thought.

Esposito's early analysis of Vico is certainly worth revisiting. Esposito is a master of finding new ways of thinking about necessity in an author. Esposito's reading of Vico raises questions that scholars have overlooked. More importantly, Esposito's reading opens up a road to connect Vico to modern political ideas. Esposito has helped to show that his own words, the last words of his text on Vico and Machiavelli, may be correct, "It ought not surprise us if, in the difficult circumstances of the bankruptcy of their ideas (*nella forbice impietosa del loro fallimento*), the political philosophers of our time seek inspiration with an increasing persistence in the thoughts of Vico even if they are often unaware of it" (*MV*, 297).

Notes

1. Giambattista Vico, *The New Science of Giambattista Vico, unabridged translation of the Third Edition*, trans. Thomas Goddard Bergin and Max Harold Fisch (Ithaca, NY: Cornell University Press, 1968). This will be cited in the text as *SN* with the paragraph number. Also Giambattista Vico, *Opere*, ed. Andrea Battistini (Milan: Arnoldo, 1990).

2. The translations of these texts are mine. The emphases are from the author.

3. Esposito gives a later gloss on this in *LT*, 72–73.

4. James Goetsch, *Vico's Axioms: The Geometry of the Human World* (New Haven, CT: Yale University Press, 1995).

5. For a discussion of these oxymorons, see Enrico Nuzzo, "Gli 'Eroi Ossimorici' di Vico" in *Tra Religione e Prudenza: La 'Filosofia Practica' di Giambattista Vico* (Roma: Edizioni di Storia e Letteratura, 2007), 211–32.

6. Giambattista Vico, *On The Most Ancient Wisdom of the Italians Unearthed from the Origins of the Latin Language*, trans. L. M. Palmer (Ithaca, NY: Cornell University Press, 1988), 45–47.

7. For an account of Vico's etymologies, see Davide del Bello, "Forgotten paths: The making of Vico's etymology," *Semiotica* 113, no. 1/2 (1997): 171–88.

8. For a classic reading of Vico as an anti-Enlightenment thinker, see Isaiah Berlin, *Vico and Herder: Two Studies in the History of Ideas* (New York: The Viking Press, 1976). For an important recent response to this interpretation of Vico, see David L. Marshall, *Vico and the Transformation of Rhetoric in Early Modern Europe* (Cambridge: Cambridge University Press, 2010), 13–21.

9. Vico himself talks about his influence in Venice in his *Autobiography*. Giambattista Vico, *The Autobiography of Giambattista Vico*, trans. Max Harold Fisch and Thomas Goddard Bergin (Ithaca, NY: Cornell University Press, 1983), 182–87. See also Christopher Drew Armstrong, "Myth and the New Science: Vico, Tiepolo, and the Language of the 'Optimates,'" *The Art Bulletin* 87, no. 4 (Dec 2005): 643–63.

10. In chapter 2 of Book 1 of the *Discourses*, Machiavelli suggests how all the three traditional forms of government may all fall into corruption (Niccolò Machiavelli, *The Discourses*, in *Selected Political Writings*, trans. David Wootton (Indianapolis, IN: Hackett Publishing, 1994), 87–93). Rousseau presents an account of the downfall of humanity in his second *Discourse* (Jean-Jacques Rousseau, *Discourse on the Origin of Inequality*, trans. Judith R. Bush, Roger D. Masters, Christopher Kelly, and Terrence Marshall (Hanover, NH: University Press of New England, 1992), 63–67).

11. For a good account of the *sensus communis* and its relation to the hermeneutic tradition, see John D. Schaeffer, *Sensus communis: Vico, Rhetoric, and the Limits of Relativism* (Durham, NC: Duke University Press, 1990).

12. Hans-Georg Gadamer, *Truth and Method*, 2nd rev. ed., trans. Joel Weinsheimer and Donald G. Marshall (New York: Crossroad, 1989), 19–24.

13. "*Il sapere practico, come quello che procede per differentiam, va sempre dal concreto all'astratto, o meglio si ferma, si riduce, si limita al concreto, consapevole delle finitezza che—direbbe Gadamer—lo estranea a se stesso; il sapere razionale, invece, presumendo ed incautamente sperimentando una via che dall'astratto porti al concreto, resta bloccato, pietrificato salla sua medesima presunzione, nel filo spinato dell'astratto e sconta così le sue inverificate ed inverificabili pretese di organicità ed assolutezza*" (*MV*, 170).

14. This view may be found in Benedetto Croce who, in making Vico Hegelian, emphasized the separation of the stages of Vico's history. Leon Pompa traced Vico's philosophy of history through analysis of causes moving civilization through distinct stages. Benedetto Croce, *The Philosophy of Giambattista Vico*, trans. R. G. Collingwood (New Brunswick: Transaction Publishers, 2002). Leon Pompa, *Vico: A Study of the New Science* 2nd ed. (Cambridge: Cambridge University, NJ Press, 1990).

15. This certainly true of anyone focusing Vico's account of the development of human institutions, as mentioned in the earlier footnote, or on anyone investigating Vico's account of poetic wisdom, such as Verene, Cantelli or Chabot. Donald Phillip Verene, *Vico's Science of Imagination* (Ithaca, NY: Cornell University Press, 1981); Gianfranco Cantelli, *Mente, corpo e linguaggio. Saggio sull'interpretazione vichiana del mito* (Florence: Sansoni, 1986); Jacques Chabot, *Giambattista Vico: La Raison du Mythe*, (Aux-en-Provence: Sarl Edisud, 2005).

16. Esposito writes, "*In genesi dell'organizzazione politica si allaccia indissolabilmente con la pubblicizzazione della violenza: l'organizzazione politica è l'organizazzione pubblica della violenza, la razionalità della formazione economico sociale è la razionalizzazione della violenza economico-sociale*" (*VR*, 109). Note also that while violence is de-emphasized in Esposito's account of Vico in his book on Machiavelli, force plays a significant role (*MV*, 264–66).

17. Jean-Jacques Rousseau, *Confessions and Correspondance, Including the Letters to Malesherbes*, trans. Christopher Kelly (Hanover, NH: University Press of New England, 1995), 3.

18. Vico directly claims that his *Autobiography* is an attempt to apply his philosophical project to his own life. He writes, "he wrote it [the *Autobiography*] as a philosopher, meditating the causes, natural and moral, and the occasions of fortune; why even from childhood he had felt an inclination for certain studies and an aversion from others; what opportunities and obstacles had advanced or retarded his progress; and lastly the effect of his own exertions in the right directions, which were destined later to bear fruit in those reflections on which

he built his final work, the *New Science*, which was to demonstrate that his intellectual life was bound to have been such as it was and not otherwise" (Vico, *Autobiography*, 182). See Donald Phillip Verene, *The New Art of Autobiography, An Essay on the Life of Giambattista Vico Written by Himself* (Oxford: Oxford University Press, 1991).

19. For an extended account of Jansenism in Vico's Naples, see Nicola Badaloni, *Introduzione a G. B. Vico* (Milano: Feltrinelli Editore, 1961), 229–58. For an account of the influence of the French tradition of the introspective philosopher, see Nuzzo, *Tra Religione e Prudenza*, 19–118.

20. Benedetto Croce, advancing his reduction of Vico to Hegel, famously claimed that Vico did not descend into practical wisdom. Benedetto Croce, *The Philosophy of Giambattista Vico*, trans. R. G. Collingwood (New Brunswick, NJ: Transaction Publishers, 2002), 227. With a sufficient understanding of the political situation in Naples at the time, it becomes obvious that Croce's reading is incorrect. Giuseppe Giarrizzo, for example, develops a compelling account to suggest that Vico's account of the past is based on his understanding of the present, which then leads to a developed political agenda. Giuseppe Giarrizzo, *Vico La Politica e La Storia* (Napoli: Guida Editori, 1981), 118–20. This view also ends up being quite different from Esposito's.

21. *"È vero che l'utile non constituisce ancora il punto d'appoggio per l'onesto, la sua molla interna, eppure dove la parabola dell'onesta si conclude, dove il suo arco misurabile si arresta, l'utile inevitabilmente riemerge"* (*VR*, 49).

22. *"L'ideologia non è solo il laccio provvidenziale che ci lega all'esistenza in quanto esseri pesanti, ma è addirittura un polo magnetico che, generato al di là del momento individuale dal lavoro colletivo degli uomini, ne fonda riproduce i rapporti; solo nella lente magica dell'ideologia può specchiarsi la civilità, può cioè—come si esprime Vico in un passo davvero machiavellico—istituirsi la vita umana come vita civile, come vita sociale"* (*VR*, 62).

23. Rousseau describes the separation of the individuals from the Sovereign in this way, "The right that the social compact gives the Sovereign over the subjects does not exceed, as I have said, the limits of public utility. The subjects, therefore, do not have to account for their opinions to the Sovereign, except insofar as these opinions matter to the community." Then later he describes the civil religion this way, "There is, therefore, a purely civil profession of faith, the articles of which are for the Sovereign to establish, not exactly as Religious dogmas, but as sentiments of sociability without which it is impossible to be a good Citizen or faithful subject." Jean-Jacques Rousseau, *Social Contract*, trans. Judith R. Bush, Roger D. Masters, and Christopher Kelly, (Hanover, NH: University Press of New England, 1994), 222.

24. Vico writes, "Now in free commonwealths if a popular man is to become monarch the people must take his side, and for that reason monarchies are by nature popularly governed: first through the laws by which the monarchs seek

to make their subjects all equal; then by that property of monarchies whereby sovereigns humble the powerful and thus keep the masses safe and free from their oppressions; further by that other property of keeping the multitude satisfied and content as regards the necessities of life and the enjoyment of natural liberty . . ." (*NS* 1008).

25. "*Per essere potente, il politico deve—ecco il suo destino epocale—porsi fuori dalla storia. Lo Stato ha una storia (anche se arcaica, improduttiva), è storia; il politico è ciò che la fonda, la attraversa, ma non la vive: è fuori dalla representatione storica, storicamente irrappresentabile come oggetto di storia*" (*MV*, 131). The emphasis is Esposito's.

26. Esposito briefly returns to this discussion of fortune and security later (see *LT*, 55).

27. "*La fortuna, in questo caso, non è un'entità positiva, reale, oggettiva: è semplicemente non virtù, ciò che manca, ciò che limita—ma dall'interno e senza forza d'urto—la virtù, ciò che la costeggia e la borda*" (*MV* 161).

28. "*non può bastare a Vico . . . distinguere una catena, ancora simultanea, di differenze (la via di Machiavelli) . . .*" (*MV* 273).

29. Vico writes at the end of his major paragraph on the barbarism of reflection: "And the few survivors in the midst of an abundance of the things necessary for life naturally become sociable and, returning to the primitive simplicity of the first world of peoples, are again religious, truthful and faithful. Thus providence brings back among them the piety, faith, and truth which are the natural foundations of justice as well as the graces and beauties of the eternal order of God" (*SN*, 1006).

30. "*L'origine della fase non è altro che la fine della precedente e visaversa. La storicizzazione è integrale, ma la possibilità del suo rischiaramento scientifico è sempre indissolubilmente legata ad un orizzonte di visibiltà capace di simultaneamente i vari tempi e la logica specifica del loro rapporto ('il limitare) . . .*" (*MV* 227).

31. For an interesting account of the information Vico was working with that shows his dedication to staying current, see Gustavo Costa, "Vico's Salti Nitri and the Origins of Pagan Civilization: The Alchemical Dimension of the *New Science*," *Rivisti di studi italiani* 10 (1992): 1–11.

2

Thinking Community

Diane Enns

"Nothing seems more appropriate today than thinking community," writes Roberto Esposito at the start of *Communitas*; nothing is more necessary or demanded, and yet nothing is further from view, "remote, repressed, and put off until later" (*CM*, 1). This is not to say that philosophers have neglected community, Esposito adds—on the contrary, it has become a prominent theme—but they distort community by translating it into a political-philosophical lexicon.[1] Recent accounts of community remain within its "unthinkability," in fact, constituting its most symptomatic expression (*CM*, 1). While community is a necessity since it is our originary condition, Esposito concludes, it is nevertheless impossible; community can never be fully realized (*TTP*).

Esposito embarked on this project of thinking community in the years leading up to George W. Bush's declaration of the war on terror. While community is inherently structured by the dichotomy of inside and outside, and thus founded on borders, the security paradigm instigated by the war on terror has intensified the preoccupation with borders into an obsession. In *Immunitas* Esposito describes this dynamic as an "immunitary" paradigm; though nothing new, it has become the most fundamental organizing principle of global politics today. He is not the only thinker to make an analogy between the body's immune system and the biopolitical operation of immunity in our times, but Esposito develops it into a rigorous account that renders immunity the linchpin around which both the real and the imaginary practices of an entire civilization have been constructed.[2]

Esposito's compelling analysis of community and the immunitary function that inhibits it inspires further thought on identity and difference, on commonality and division, and on collective responses to violence and risk. His project is ambitious: to think community (despite its unthinkability) in such a way that we avoid defining community as property or substance that joins us together, for what is held in common belongs to no one. Esposito thus follows a trajectory of recent thought on community that employs the negative—referred to as "no-thing" or nihilism—to permanently intervene in the community's tendency toward essence and the closure this essence implies. In other words, he seeks to understand how we can protect the community against too much protection. To avoid the dialectic of identity and difference, Esposito investigates the etymological origins of community, more precisely, the *munus*—gift or obligation—that we find at the heart of both community and immunity.

But we are also inspired by Esposito's project to reflect on the meaning of community life—how we live, enjoy, and desire human togetherness—because this is the affective dimension he neglects to investigate. In what follows, I will consider the implications of Esposito's account and point to the limitations of his etymologically informed narrative by considering the lived experience of community. I will argue that the formulation of a deconstructive impossibility—the nihilism Esposito evokes—is theoretically significant, but it ignores vital aspects of community life. To maintain that community is in one sense unthinkable and impossible only reaffirms it as an ideal of perfect unity or communicability. But this ideal is not community as we live it; nor is such an ideal desirable since it would constitute the very totality—pure unity—that Esposito wishes to avoid. We can think community, and not despite its unthinkability as Esposito would have it, because we live it. We can reflect on our experience of community in its fragile imperfection, as we move collectively between the enjoyment of human togetherness and the risk of conflict and violence.

Turning our attention to the desire for, and enjoyment of, community does not negate the risk of violence in the face of incommensurable difference; nor does it eliminate the risk of communities closing in around essentialized identities. In Esposito's intriguing analysis of the "tolerant immunity" with which the pregnant woman's body regards the foreign other developing inside her, we find his most promising insight into how we might think community in such a way that avoids those perils we ignore in our obsession with security. But if we are to meet the demand to think community today, "heralded by a situation that joins in a unique epochal knot the failure of all communisms with the misery of new individualisms" (*CM*, 1), we must take into account the valued elements of community that contribute to a meaningful life and motivate what Edith Stein called "the urge to unify."

Immunitary Obsessions

Even a cursory glance around us confirms that Esposito's diagnosis of the immunitary paradigm is apt—it organizes more than global politics, and has significantly shaped our understanding of the meaning of community and how we experience it. Sensitive to the ubiquitous manifestations of immunity around us, he accurately depicts our demand for protection and examines its effects. We live in a risk-averse society: at the micro level, we expect to be saved by hand sanitizer, rigorous vitamin and exercise programs, insurance and pension plans. At the global level the stakes are much higher, with attempts to guarantee immunity through preemptive military strikes, biometric border control, and the construction of walls. In the fields of disease control and information technology the virus has become the metaphor for all our nightmares (*TTP*, 60). At the desks of policymakers, protection against contamination by foreign elements—immigrants, migrants, terrorists—is the new religion. Esposito captures this operation well: "instead of adjusting the level of protection to the actual presence of risk, we are adjusting instead the perception of risk to the growing demand for protection." Risk is artificially produced in order to control it (*TTP*, 62).

Esposito analyzes both old and new forms of this narrative. Immunitary effects have always been constitutive of religion's promise of immortality, for example. Faith in God immunizes believers against the suffering caused by life's vicissitudes, existential loneliness, and the fear of death; religion, as Marx observed, is the opium of the people. The law also immunizes, protecting the individual and political body from the excesses of violence and disorder; like a vaccine, it injects into the body some element of the disease against which the body must fight. The death penalty and the police apparatus are examples of such social "vaccinations," administering small doses of the very violence against which the community is shielding itself. This is sanctioned violence with an obvious immunitary function: violent means are used to exclude violence. The police carry the instruments of death that could take lives and destroy communities in order to protect us. As Esposito puts it: law is "violence against violence in order to control violence" (*IM*, 29).

But immunity is a double-edged sword. Our bodies need immune systems, both at the individual and the social level. Religion can provide a harmless benefit for the individual and for members of a community, if it is not the basis of discrimination or extremism. Our police systems are also beneficial in an immunitary sense, in that they allow us some measure of protection under normal circumstances. Depending on where we live we might be relatively confident that our police will enforce the laws we hope will deter drunk drivers, petty thieves, and violent offenders, and come to our aid in emergencies.

If carried past a certain threshold, however, immunity grows out of proportion—the organism attacks itself and it explodes or implodes, negating life (*TTP*, 61, 62). We know how this happens in the body—diseased cells destroy their innocent neighbors in a flurry of overprotective activity. The immune system spirals out of control, turning into a self-destructive autoimmune disease. And so religion turns into religious fundamentalism: the faithful prove themselves by strapping explosives to their chests and taking as many of the enemy with them as they can. Policing turns into police brutality: the same forces whose job it is to deter theft and murder might point their weapons at protesters, and arrest, beat, or kill when called upon to do so.

Both of these examples of excessive immunity come into play in the current war on terror. Esposito explains that the conflict of this war bursts out of two immunitary obsessions: on the one hand an Islamic extremism that is determined to protect to the death what it considers to be its own religious, ethnic, and cultural purity from contamination by Western secularization, and on the other, a West that is bent on excluding the rest of the planet from sharing its own excess goods. Once these oppositions were bound together, he argues, "The entire world was convulsed by what resembled the most devastating autoimmune disease. A surplus defense with regard to elements outside the organism had turned against the organism, with potentially lethal effects" (*TTP*, 62). This has occurred within a monotheistic triangle; Christianity, Judaism, and Islam are all devoted to "the logic of the One," Esposito notes, unifying the world according to their own points of view. The war on terror reveals the immunitary paradigm in full swing (*TTP*, 63).

Whatever the history of the immunitary function, the fact that the public largely fails to call for a justification of surplus defense attests to the pervasive normalization of what we might call "immunitary monologic," especially in the realm of global politics. There is little public outcry against the use of drones to target suspected terrorists in Pakistan, Syria, or Iraq, against the walls erected to keep undesirables out, or in opposition to biometric screening and public surveillance. The rhetoric of immunity has crept into every corner of the globe, legitimizing acts of violence against real or perceived enemies in the name of security against terrorism. It works thanks to the ease with which a population can be provoked to fear any potential threat with the help of an alarmist media. Another sense of immunity enters the picture here—the immunity of the powerful from law and justice, from respect for another's humanity.

Communitas without the Common

The war on terror illuminates Esposito's keen insights on the immunitary paradigm as the organizing principle in today's global politics, but it also corroborates his point that an obsession with risk and threat "immediately liquidates contact, relationality, and being in common" (*TTP*, 59). In fact, he goes so far as to say that "immunity (or, in Latin, *immunitas*) is the opposite of *communitas*" (*TTP*, 58). Whether at the state level or in our neighborhoods, collective life is jeopardized when security interests are promoted at the exclusion of all else. Walls keep undesirables out but imprison those within; murder in the name of security might turn into murder-suicide.[3]

We do not have to look far to find evidence of the effects of too much immunity, from the suffering and ignorance caused by religious sects at war with each other to the hostility fueled by political ideologues whose purpose relies on an essentialized identity, victimhood status, and ideological conditioning. The stakes vary but the underlying referent is the same: the authenticity of one's claims to this or that identity category whether it entails an adherence to ideology or bloodline. Immunity, Esposito explains, grants a release from the debt that binds the community. Individuals are "exonerated, or relieved of that contact, which threatens their identity, exposing them to possible conflict with their neighbor, exposing them to the contagion of the relation with others" (*CM*, 13).

What would Esposito have us do? If immunity configures a community inevitably "walled in within itself and thus separated from the outside" (*CM*, 16) we need to ask whether other forms of community are possible.

Can a community remain open to what lies beyond its borders, and to the risk and threat of what might be hidden within? Esposito articulates two responses to what he believes is the central problem in thinking community: defining the common as that which unites the property of each of its members—"they have in common what is most properly their own; they are the owners of what is common to them all" (*CM*, 3).

The first response we find in *Communitas*, Esposito's exploration of the historical, philosophical, etymological meaning of community. He turns to etymology for help in counteracting the persistent historical myth that we must reproduce community in its "originary essence" (*CM*, 16), an "affirmative entification" from which we must withdraw by observing the lack at the heart of community. This is his objective in *Communitas*, to reject the fullness of subjective substance, identity, fusion, intersubjectivity, and expose the void or lack that ceaselessly interrupts this fullness. This entails questioning the common assumptions about community that have led to its acceptance as a property that belongs to subjects and joins them together, a predicate that qualifies them as belonging to the same totality or an additional substance added to their nature as subjects. That community is considered most properly our own is an obvious paradox, Esposito points out, for what is most common cannot be owned by an individual (*CM*, 3).

Esposito finds an escape from this dialectic in the "*munus*," Latin for gift, debt, or obligation that lies at the heart of both *communitas* and *immunitas* (*CM*, 5–6). Through an etymological excursion he tells us that community does not mean owning a property, but giving a gift. This is a particular kind of gift, distinguished by its obligatory character. "In short, this is the gift that one gives because one must give and because one cannot not give" (*CM*, 5). The ancient meaning of *communis*, he tells us, had to be "he who shares an office," a "burden," or a "task" and *communitas* arises then as the totality of persons united by an obligation or debt, an "I owe you" rather than an "you owe me" (*CM*, 6). This debt presupposes a lack, since members of a community are expropriated of the initial property of their own subjectivity (*CM*, 7). In a community subjects do not find a principle of identification, they only find the void that is this lack. The community is not a body in which individuals are parts of a larger individual; community is not intersubjectivity or *inter-esse* (*CM*, 7). Nor is there recognition of a collective identity in this schema. Community is a shared nothingness or no-thingness.

Esposito's analysis highlights the negative construction favored by a generation of European thinkers grappling with the practical and philosophical legacies of fascism and totalitarianism. Suspicious of gathering or coinciding, of unity and commonality, they seek to intervene in any positive sense of community; all are critical of belonging, identification, and the exclusivity that appears to be their necessary outcome.[4] Alphonso Lingis is perhaps most eloquent on this point when he writes: "When individuals associate, they identify those outside their agreements as barbarians and monsters; the effective recognition of common humanity extends as the exchange of messages, resources, and services with outsiders establishes an agreement with reciprocal commitments."[5] Like Esposito, he seeks to displace the focus on belonging or identification with debt or obligation. Even kinship, Lingis tells us, is not about recognizing family resemblances, but obligation.[6] This is not suspicion of totalizing gestures only, but also of a rational, communitarian approach to common life that remains a positive affirmation of intersubjectivity. To argue that community is more truly among "those who have nothing in common," an encounter with the other, with "the intruder," is intended to disrupt an insulated, essentialized whole. The "other" community that Lingis posits, for example, is not realized in producing something in common but in exposing itself to the stranger who has nothing in common. It is the demand of this stranger that constitutes community.

The negative is introduced by these thinkers in order to address the question of absolute difference or incommensurability. Our limit cases—face to face with a member of the Taliban or ISIS, or the armed person claiming "Stand Your Ground" immunity, for example—defy our positive constructions of rational communities based on liberal values of tolerance, good will, and communication. But if the only problem with intersubjectivity appears to be the risk of excluding the other—in other words, the risk of immunity gone to excess—is this a risk we need to accept? Does community have to close in on itself and perform the imprisonment or murder-suicide of excessive immunity? Evidently, Esposito does not think so, as we will see.

Esposito's discussion of nihilism and community contributes to critiques of identity at a theoretical level but widens the gap between thinking community (as unthinkable) and lived experience. This is most pronounced in the "Appendix" of *Communitas* where he reiterates his criticism of the typical opposition made between nihilism and community by "communal, communitarian, and communicative conceptions that for more than a century have found community to be the only shelter from the devastating power of nothing that goes completely unchecked in modern society" (*CM*, 135). Nihilism is not an insurmountable obstacle to community, Esposito insists, "but instead the occasion for a new way of thinking community." While not the same, nihilism and community cross each other; they have in common no-thing (*CM*, 137). This is the outcome of his etymological investigation, for what the *munus* demonstrates is "the categorical distance" that community enjoys vis-à-vis every conception of collectively owned property or common identity. Given the debt or gift at the core of community, its members share an expropriation of their own essence (*CM*, 138). They are no longer identical with themselves but exposed to others, forced to open their own individual boundaries. This means, Esposito concludes, that the community always consists of others, as the subject is "no longer the 'same,'" the community is never itself, inhabited by an absence of subjectivity, identity, and property. It is thus defined by its "non" or lack, its no-thing (*CM*, 138).

The implications of this lack are made explicit in the following passage:

> Indeed, [community] is the departure itself toward that which doesn't belong to us and can never belong to us. For this reason *communitas* is utterly incapable of producing effects of commonality, of association [*accomunamento*], and of communion. It doesn't keep us warm, and it doesn't protect us; on the contrary, it exposes us to the most extreme of risks: that of losing, along with our individuality, the borders that guarantee its inviolability with respect to the other; of suddenly falling into the nothing of the thing [*niente della cosa*]. (*CM*, 140)

This passage strikes me as a rather barren depiction of community. To say that the no-thing separates members of a community from themselves, or that there are only others in this configuration—no selves, only lack—does not sound like a community in which we might take pleasure. Why would we desire what merely burdens us with duty and sacrifice? That the other expropriates the essence of the self and creates a lack assumes that we begin with an essence that must be relinquished in the debt to others. But we are not property in ourselves—if we do not own our essence then there can be no expropriation. If we acknowledge that a self is always-already constituted by its other—that there is an individual but no pure individual *essence* since the self is formed along with others—then there is no lack. If we begin from the premise that the other gives me to myself, the direction of the gift is reversed. It is not a question of replacing "You owe me" with

"I owe you" but "I give to you and in giving to you I am given back to myself." There is gift in this formulation, but no demand or debt.

Esposito is right to say that the common is "not possession, property, or appropriation" only if this means an absolute possession. The lack refers again to an ideal we can never attain—a "coming community," to use Giorgio Agamben's formulation, that may have little to do with the commonalities we enjoy as members of communities that are always contingent, vulnerable, open to risk. Our actual communities seem marked by *fullness* rather than lack, by *sharing* rather than paradoxically owning what is properly our own. Community is the *inter-esse*. Despite Esposito's vehement claims to the contrary, in his second, arguably more ambivalent, response to the central problem of community we discover a promising articulation of the intersubjectivity of community, one that does not deny the demand for protection or the risk that threatens it.

Immune Tolerance

We can imagine a philosophy of immunity that does not deny its inherent contradiction, Esposito writes, yet reverses the semantics in the direction of community (*IM*, 165). Immunity and community would then become indistinguishable, or we might prefer to say, reversible. He turns to human biology for an example. In the case of a pregnant woman's acceptance of the foreign body growing inside her own, we find a kind of immunity that is not self-destructive; the pregnant woman has an immune system that welcomes a stranger.

Relying on Alfred Tauber's work on immune tolerance, Esposito begins his discussion by asking how the fetus, "encoded as 'other' based on all normal immunological criteria," is tolerated by maternal antibodies (*IM*, 169). The mother's immunity mechanism works "on a double front" controlling the fetus as well as itself; in essence, immunizing itself from an excess of immunization (*IM*, 170). This is in keeping with Tauber's claim that the immune system does not only maintain organic integrity, its main function is to define the identity of the subject in its dynamic interaction with its environment. The body is not a closed unit constituted by "impassable borders," according to Tauber, but an ecosystem that has evolved over time into a "social community." There is still an "outside" of the body-self, but rather than attack and destroy it, the immune system is in constant relation to this outside, "a *process* that always involves an open system of self-definition that consistently produces self and other" (*IM*, 169). The self is not a constant then, but a principle of action; "Its boundaries do not lock it up inside a closed world; on the contrary, they create its margin, a delicate and problematic one to be sure, but still permeable in its relationship with that which, while still located outside it, from the beginning traverses and alters it" (*IM*, 169).

In pregnancy, the mother's antibodies permit the embryo to survive by blocking the signals that indicate the embryo is foreign. Her body essentially immunizes itself from an excess of immunization. This is not a failure of what we might call the immune system's "security apparatus" but part of the system's activities. Important to note as well is the fact that the operation depends on sufficient foreignness in the father, for too much similarity would result in miscarriage. Esposito deduces from this that "only as a stranger can the child become 'proper.' " Indeed, he stresses that the myth of symbiotic unity between mother and fetus is exaggerated; in fact, they are engaged in "a furious battle" (*IM*, 170). He elaborates:

> The mother is pitted against the child and the child against the mother, and yet what results from this conflict is the spark of life. Contrary to the metaphor of a fight to the death, what takes place in the mother's womb is a fight "to life," proving that difference and conflict are not necessarily destructive. Indeed, just as the attack of the mother protects the child, the child's attack can also save the mother from her self-injurious tendencies—which explains why autoimmune diseases undergo regression during pregnancy. (*IM*, 171)

This is the point at which *immunitas* becomes indistinguishable from *communitas*, when nothing remains of the incompatibility between self and other, since inside and outside, proper and common mingle; immunity and community intersect (*IM*, 171). But we could just as easily amend the above passage to emphasize the mother's welcoming of the stranger. The "fight to life" also proves that sameness or unity is not destructive. The infant is being nourished and protected *within* the mother's body, an example par excellence of life-giving security.

As if anticipating this observation, Esposito comes to the heart of the matter in the remarkable passage that follows this discussion:

> The equilibrium of the immune system is not the result of defensive mobilization against something other than self, but the joining line, or the point of convergence, between two divergent series. It is not governed by the primacy of the same over the like and the like over the different, but by the continuously changing principle of their relationship. In this sense, *nothing is more inherently dedicated to communication than the immune system*. Its quality is not measured by its ability to provide protection from a foreign agent, but from the complexity of the response that it provokes: each differential element absorbed from the outside does nothing but expand and enrich the range of its internal potential. . . . Before any other transformation, each body is already exposed to the need for its own exposure. (*IM*, 174; emphasis added)

Which comes first, the welcoming of the other or the battle against it? We need not choose one over the other; the privileged moment is rather the constant negotiation that is their relation. Indeed, if we extend this biological example by considering the pregnant woman's relation to her child after the event of birth we see how "furious battle," but also *furious protection*, continues between mother and child. The infant embarks on a long and ambivalent journey of separation, and the mother on a corresponding journey of loss. The "foreign" encroaches slowly and always painfully as parent and child negotiate borders. If separation and difference are not tolerated, both will suffer. The child and the mother are equally capable of devouring each other; they risk mutual destruction.

The analogy Esposito makes here is brilliant when applied to community life in its political dimensions, especially when community is forced to confront the absolute limits of the dialectic of identity and difference he has analyzed in such depth. For the work of maintaining the permeability of borders is severely constrained when those bent on death and destruction (whether inside or outside of the gates) must be kept from harming others. Immune tolerance would mean avoiding military solutions and seeking the points of convergence in any political conflict; above all, it demands communication. The results are not measured according to how well such an approach provides security, but according to "the complexity of the response it provokes." Destruction provokes the simplest response: more destruction. Communication provokes any number of responses (the risk of violence among them), but the most complex of these might be understanding and recognition, the starting points for dealing with conflict. In the face of this limit, the community must, sometimes at great risk, remain exposed to the need for its own exposure.

The Common

Esposito has given us two responses to the dialectic of identity and difference constitutive of community. In the first, lack or no-thing disrupts the tendency of borders to close in on themselves, and severs the relation of the community to what is held in common; in the second, borders are permeable, exposed to the foreign, negotiating conflict with this other in order to coexist. The coexistence of the pregnant woman and her unborn infant—the convergence and divergence of their biological borders—beautifully illustrates the *inter-esse* or between-being of human existence. But in neither approach does Esposito take into account our desire to experience this between, to be together with other bodies and minds—to belong and to share—nor does he address the affective dimensions of communal existence.

We might then ask: Where is the community in Esposito's analysis? The pregnant woman and her fetus are strangely empty figures in his discussion of immunology; their *inter-esse* reduced to a biological exchange. The mother's feelings of longing, love, indifference, or anger toward her unborn child do not factor into the negotiation of borders between them. Rather than lack, we might reflect on the *fullness* of the pregnant woman's body as a metaphorical description of a meaningful community life. In other words, to complement Esposito's astute analysis of the dialectic of identity and difference as it pertains to borders and immunity, we need to flesh out more fully the *common* of community, and this necessitates an inquiry into the shared affective, sensual life of its members. He might protest at this line of thinking, which appears to construct yet another positive entification of community but, like case of the pregnant woman who defies any easy inside/outside dialectic, we do not experience communities only as enclosed, as hostile to difference.

If we examine our own experiences of community it seems apparent that despite Esposito's claim to the contrary, what we hold in common forms the basis of our desire for and pleasure in belonging to groups; we crave identifications with a wide spectrum of collectives formed around families, friends, neighbors, and shared interests—cultural, national, religious, intellectual and so forth. Notwithstanding its etymology, I question how much community has to do with obligation and duty, with a gift "that does not belong to the subject, indeed that weakens [*reduce*] the subject and that hollows him out through a never-ending obligation, one that prescribes what is prohibited and prohibits what is prescribed" (*CM*, 17). We participate in communities because of our enjoyment in sharing a common world, and this acknowledgment opens the discussion onto a wider social terrain. Our social relations do not "hollow" us out, but fulfill us, and in so doing, return us to ourselves. Just as friends or loved ones give shape and definition to who we are, we do not lose ourselves within a community, but find ourselves. We act and interact within the community for the benefit of the community, not out of duty, but out of pleasure in shared pursuits and conversation.[7] We make communities, but they also make us.

We also participate in community because we cannot do otherwise without an effort. We simply find ourselves in any number collective configurations, whether chosen or arbitrary. Not all communities are based on uncompromising, essentialized identifications; we also form loose or arbitrary communities, simply out of the need for human contact and for "home," defined in a broader sense than a personal dwelling. Étienne Balibar refers to "communities of fate," a designation he borrows from Herman van Gunsteren, meaning "*any place where individuals and groups belong*," wherever these groups happen to have been thrown together by history. They already include difference and conflict; their interests and ideals "cannot spontaneously converge, but also cannot completely diverge without risking *mutual destruction* (or *common elimination* by external forces).[8] This is a community of chance whose members, Balibar notes, are at the same time radically foreign to one another and unable to survive without each other.[9] Recall the furious battle (and furious protection) between pregnant woman and fetus, or mother and offspring; what we find here is the persistent movement from divergence to convergence and back again. The element of fate or chance defies the affirmative entification about which Esposito worries. A common is not owned like property, but *shared*. In a sense the common owns us—we *belong* to it. These terms—sharing and belonging—also dislodge the association of the common with what is owned.

The Urge to Unify

The desire for community arises out of our orientation to the world of other living beings. Like the "clinamen" Jean-Luc Nancy evokes to describe the individual's inclination toward others,[10] we are always busy creating community, orienting our lives like swerving atoms around multiple relationships, from our gatherings at work, in neighborhoods, in clubs or political associations, with small groups of friends or family members. We are drawn to those with whom we have something in common—this common, rather than difference, gathers us into these communities and keeps us committed to them. The foreign also attracts us—we are fascinated by what is other to us, because it draws us into new worlds, beyond our own cares, thoughts, and experiences. Inevitably, however, we seek out what is common between us—even between strangers—raising the question: are we so radically foreign to one another? We may not share a language, but we smile and nod over a universally shared experience—observing the antics of children, for example—or we shed tears in the face of a stranger's suffering. When sworn enemies shake hands we cannot resist the feeling of triumph that comes with mutual understanding. Drawing attention to what we hold in common rather than the impossibility or no-thingness of community does not mean that we deny or neglect conflict and the possibility of violence, but that we do not dismiss all unity as destructive of difference and pluralism. The warmth of congenial relations and the mutual affirmation of our selves through communication enable us to survive the differences that threaten to undermine the warmth of these relations. This gives us another example of immune tolerance and the conditions required to negotiate and overcome conflict.

While community relationships are not the same as friendships, the distinction hinges on degrees of intensity and intimacy. Our friends do not necessarily constitute a group of any kind. Community provides us with less intimate relationships for the simple reason that Aristotle noted: just as we cannot be in love with many people at once, we cannot have too many friends, for it is very hard to "acquire some experience of the other person and become familiar with him."[11] We need looser connections with others, to help us establish our own familiar worlds, places where we uniquely belong. Our social fields range from the intensity of love relationships to somewhat less intimate friendships, and so on to neighbors and acquaintances who make up our communities, to the strangers bound only by geographical borders of cities or states, and ultimately the globe.

What holds any of these together is always whatever we have in common, whether it is the shared world of lovers, religious values, a cultural heritage, or even the random events of a particular geographical location that we must survive together—heat, tropical storms, winter ice and snow. At the global level we must survive as one planet in a universe—a conglomerate of interdependent species and ecological systems.

We actively seek out what is familiar to us in order to create a home, and in the process, establish borders and strangers. We step outside our doors and take pleasure in greeting neighbors, in establishing a rapport, however superficial, with our local grocers or mail carriers. This is by no means a universal pleasure, and, I suspect, increasingly difficult in a technologically sophisticated society such as ours. But when community is absent in proximity, it is sought virtually. This search is about a fundamental need to belong to a place and to create habitats with those around us. Such a wide spectrum of experiences of belonging and sharing is not reducible to a narrowly defined positive entification.

Community is not only structured on need; we *value* the experience of a shared existence. Here I employ Edith Stein's expression, the "urge to unify," to capture our yearning for the value that community provides. Stein writes that the greater the community is valued by an individual, and the more she depends on it, the more value the community will have.[12] The value that is realized in community is not only the result of the modifying effect on the individual person in the presence of others, but also the release from a loneliness Stein calls "natural" (*PP*, 273). Understanding the value of collective life helps explain our continued fascination with the social and political dynamics of community, their persistence in the face of danger or trauma, and their stubborn power once they become closed and cemented.

Although we might call on fascism and totalitarianism to exemplify this power, Stein reflects on the positive manifestations of what she calls the life force or "lifepower" (*Lebenskraft*) of the community (*PP*, 197). An animating force, it sweeps individuals together into a current of mutual feeling that becomes "their common property" (*PP*, 189). A grieving or suffering community, for example, would experience this unifying current as something larger than that experienced by an individual. When a conquered people collectively suffers, Stein explains, the experience is so vast that the individual person "stands before it as before something immense and incomprehensible" (*PP*, 143). Even if we could gather within us all of the individual experiences of a particular event, we would not capture the full communal experience; the total "experiential current" is more than the sum of its parts (*PP*, 144).

I find Stein's notion of the life force that animates and motivates community life—a moving current rather like Heraclitus's river, both same and different—one of her most compelling ideas. This power draws us to others, urges us toward union and commitment regardless of risk (*PP*, 270). There are degrees of intensity in this urge to unify, from love bonds to looser associations, from family to nation (*PP*, 270); it is a dimension of human togetherness that we must take seriously in any philosophical exploration of community. There is a soul-like quality to this life force, as though the community has an inner life bound together by individuals open to one another and to the animating power of the experiential current of community life (*PP*, 273). To speak of a community's soul is to suggest we carry the community within us; as though it were our center of gravity (*PP*, 275).

This brief phenomenological description of the lived experience of community speaks to the wider affective (social and psychological) aspects of unity or intersubjectivity that Esposito's account neglects. He does not sufficiently demonstrate why the unity or commonality of subjects is more dangerous than lack or nihilism. Indeed, if we looked for answers among those whose communities have been destroyed by the violence of surplus immunity (or not enough) we might find that communication, but also the longing to experience again the value and meaning of sharing and belonging enable the first steps into a future. When all semblances of political commitment and justice disappear, as occurs during and after violent conflict, there is still the human need to eat together, construct shelters, and converse. People start talking to one another when they have practical needs to meet, and common projects, like rebuilding the institutions and infrastructures of destroyed towns and villages. In the aftermath of violent conflict, remaking worlds that have been catastrophically unmade is not about coming to love one's neighbors, or highlighting one's differences, but about building something together. This does not establish a world without conflict, but a world in which habits are instilled in the community—new (or very old) ways of dealing with conflict that are not destructive. I would venture that no one engaged in this challenge would suggest that it is difference that keeps a community together. What we hear, instead, is that *under certain conditions* differences are tolerated, for the most part ignored, accepted with a sense of humor, because they do not matter as much as the community life among neighbors with its comforts of shared coffee and conversation. The value of community, our desire for unity and the force that brings us together, is already a form of immune tolerance.

Conclusion

To inquire into the lived experience of community life does not necessarily lead to an essentialist construction of community, or to a rational communitarian approach to human togetherness. That the *inter-esse* is defined by sharing and belonging, as well as the convergence and divergence across permeable and moving borders, is already implied in Esposito's discussion of the immune tolerance we find in the pregnant woman.

Esposito might sympathize with my phenomenological description of community, but it does not seem to enter into his analysis except in isolated moments. For example, in a reference to Machiavelli's most "explosive element"—the reversal of a negative view of violence and confirmation of its necessity for any new constitution—Esposito asks us to consider the other side of the coin: for initial violence to have ordering effects rather than destructive ones, it must be preceded and shaped by instances of virtue and right training, which derive in turn from wise laws (*IM*, 35). Could this not also refer to the kind of immune tolerance—manifest in social or political habits—that we require to avoid destructive violence, such as negotiation, communication, trust, and exposure to what is other to us; in other words, community as exposure to exposure? What comes before the violence will determine how a community is able to recover in its aftermath.

Perhaps this is also what Esposito hints at when, at the end of his discussion of nihilism in *Communitas*, he points to a new direction. From this new vantage point, we understand the "no-thing" as "the loss of mastery with regard to the complex meaning of experience" (*CM*, 149). He elaborates:

> Community refers to the singular and plural characteristic of an existence free from every meaning that is presumed, imposed, or postponed; of a world reduced to itself that is capable of simply being what it is: a planetary world without direction, without any cardinal points. In other words, a nothing-other-than-world. It is this nothing held in common that is the world

> that joins us [*accomunarci*] in the condition of exposure to the most unyielding absence of meaning and simultaneously to that opening to a meaning that still remains unthought. (*CM*, 149)

There appears to be content rather than lack in this no-thing—the world itself, a nothing-other-than-world that joins us. Or perhaps, although Esposito might not go so far, *worlds*. In other words, communities.

By exploring the meaning of community as we might experience it, I have argued that it is never difference itself that holds us together, but what we find in common with others. In his emphasis on the *munus*, Esposito neglects the affective dimensions of community—desire, pleasure, love, and the longing for meaning—essential elements of immune tolerance that we find valuable. Difference plays another role in the protection against too much protection. Our constant exposure to what is other than and beyond ourselves is what protects us from political monotheism or monoculture. Communities must negotiate difference and commonality in an endless process that always risks the violence that separates or the violence that consumes. Immune tolerance requires communication, a point Esposito concedes when he states: "Nothing is more inherently dedicated to communication than the immune system," but does not pursue (*IM*, 174).

This is not always the answer we like, for communication and congeniality—the warmth of community Esposito seeks to dismiss—and the vulnerability and risk they imply, are always context-specific and slow. But we know that communities must live with risk. The question is how to meet it, how to discern, like the pregnant woman's antibodies, when we might welcome the stranger, or expel him. This requires much more skill and foresight than militarized borders or preemptive strikes. The risk of death and violation is unavoidable, but the alternative—to be imprisoned in our own fear and risk the murder-suicide of surplus immunity—is worse.

Notes

1. Esposito refers in a note to two German works that highlight "the ambiguous 'return' of community," namely *Renaissance der Gemeinschaft? Stabile Theorie und neue Theoreme*, ed. Carsten Schülter (Berlin: Duncker & Humblot, 1990); and *Gemeinschaft und Gerechtigkeit*, ed. Micha Brumlik and Hauke Brunkhors (Frankfurt: Fishcer Taschenbuch Verlag, 1993) (*CM*, 1, note 2).

2. See for example, Jacques Derrida's use of autoimmunity in Giovanna Borradori, *Philosophy in a Time of Terror: Dialogues with Jürgen Habermas and*

Jacques Derrida, trans. Pascale-Anne Brault and Michael Naas (Chicago: University of Chicago Press, 2004); and Jacques Derrida, *Rogues: Two Essays on Reason*, trans. Pascale-Anne Brault and Michael Naas (Stanford, CA: Stanford University Press, 2005).

3. I am indebted to David Grossman for this insight. In a discussion of the effects of excessive security on Jewish Israeli citizens, he writes: "I came to grasp the high price we were paying for willingly giving up on parts of our soul—a price no less painful than giving up land. I knew that we were not killing only the Palestinians, and I asked why we were continuing to accept not just the murder, but the suicide too" (*Writing in the Dark: Essays on Literature and Politics*, trans. Jessica Cohen (New York: Farrar, Straus and Giroux, 2008), 25–26).

4. See, in particular, Jean-Luc Nancy, *The Inoperative Community*, trans. Peter Connor et al. (Minneapolis: University of Minnesota Press, 1991); Giorgio Agamben, *The Coming Community*, trans. Michael Hardt (Minneapolis: University of Minnesota Press, 1993); and Alphonso Lingis, *The Community of Those Who Have Nothing in Common* (Bloomington: Indiana University Press, 1994).

5. Lingis, *The Community of Those Who Have Nothing in Common*, 157.

6. Ibid., 156.

7. I am not suggesting that we do not perform duties or make sacrifices unwillingly in our various communities, but that duty and sacrifice are not constitutive of the meaning of community. Not all aspects of collective life are enjoyable or pleasurable, but even suffering in solidarity can provide meaning and fulfillment.

8. Étienne Balibar, *We, The People of Europe? Reflections on Transnational Citizenship*, trans. James Swenson (Princeton, NJ: Princeton University Press, 2004), 132.

9. Étienne Balibar, *Equaliberty: Political Essays*, trans. James Ingram (Durham, NC: Duke University Press, 2014), 144.

10. Nancy, *The Inoperative Community*, 3–4.

11. Aristotle, *Nicomachean Ethics*, trans. W.D. Ross, Book VIII, 6.

12. Edith Stein, *Philosophy of Psychology and the Humanities*, ed. Marianne Sawicki, trans. Mary Catharine Baseheart and Marianne Sawicki (Washington, DC: ICS Publications, Institute of Carmelite Studies, 2000), 272. Hereafter cited in the text as *PP*.

3

Debt and the Proper in Agamben and Esposito[1]

Greg Bird

Recently a debate has emerged about *la differenza italiana*, even *le differenze italiane*. Two conferences were held, the first in Paris (January 2014) and the second in Naples (May 2014), where rigorous debates ensued concerning elevation of radical Italian philosophy as a new "Italian Thought." There is a growing consensus that what makes Italian Thought different is its position of permanent opposition toward state and biopolitical forms of power, grounded engagement in the critique of capitalism, and an internal disposition toward productive conflict.[2] Antonio Negri claims that what distinguishes radical Italian philosophy is its constituent affirmation of difference, which does not lead to separation and isolation, but to resistance and creative transformation.[3] Whereas in *Living Thought* Roberto Esposito argues that Italian Thought is characterized by the immanentization of antagonism because at its core it is a theory of conflict and struggle. In his wide-ranging account of the history of Italian Thought, Dario Gentile maintains that what distinguishes it from other traditions is its foundational *soggettività antagonista* (antagonistic subjectivity). Drawing from the etymology of *sinistra* (left), he argues that the Italian left has developed a tradition of *sinisteritas*. What marks the *differenza italiana* is the internal decision to remain in a position of opposition to sovereign power. This conflictive orientation not only directly confronts sovereign power, but it also attacks it at its core by refusing to seize and thus lay claim to a sovereign position.

Most of this discussion has focused on whether contemporary Italian political philosophy is geophilosophically different enough to warrant the moniker "Italian Thought." Some have questioned whether the internal differences amongst Italian thinkers are too vast and de-territorialized to legitimate the new coinage.[4] It is not my intention to enter into the minutiae of this debate or to examine more general ontological questions; instead, I will use this discussion to examine the relationship between Giorgio Agamben and Esposito's philosophies. Esposito is obviously one of the main proponents of Italian Thought. His book *Living Thought* and many of his recent publications, interviews, talks, and university classes, consider the geophilosophical, yet open and de-territorialized, importance of contemporary Italian Thought within broader geopolitical discourse. Agamben, however, is more ambiguous. He tends not to cite his contemporary Italian thinkers and only rarely does he give a nudge to his geophilosophical predecessors or the Italian political movements that have influenced his philosophy. Given the various characterizations of Italian Thought, it would be hard to make the case that he is an Italian Theorist. For example, Esposito has recently stated that "Italian Thought" is differentiated from the French and German traditions on the grounds that "German Philosophy" is a philosophy of "negation," "French Theory" is a theory of "neutralization"; yet, Italian Thought is "active, productive," and thus "affirmative."[5] This affirmative declension is found in three contradictory pairings: the common and the immune; potentiality (*potenza*) and power (*potere*); and, conflict and neutralization.[6] In most of his work, Agamben does prioritize the common over the immune and, of course, potentiality (or impotentiality) over power, but most read him as a philosopher who prefers to neutralize conflict, which in Esposito's categorization would make him more a French than Italian Theorist.

One obvious symmetry in both philosophers is their investment in assessing how the economic has come to dominate, even erase, the political. Both have dedicated much time to critically examining how the administrative paradigm of economic theology has gradually superseded the paradigm of sovereign transcendence that once defined the political in theological terms. From their more analytical works on this topic—*The State of Exception* and *The Kingdom and the Glory*[7] for Agamben, and the now translated *Categories of the Impolitical* and *Two* by Esposito—to their applied analyses of the biopolitical governance and regulation of life—Agamben's *Homo Sacer* series and Esposito's biopolitical trilogy, amongst other works—we will surely witness a flurry of commentaries concerning their parallel, yet not symbiotic, theoretical trajectories.

The focus of this paper is different. I examine how Agamben and Esposito address the problem of community in relation to the dialectic of alienation and appropriation. *The Coming Community* and *Communitas* must be placed in a much broader group of philosophers beyond the borders of Italy and Italian Thought. These texts belong to *fin de siècle* continental debate about community. In the first part of this paper, I briefly set up this problematic by examining this dialectic in broad terms. In the second part, I focus on how each uses a radically reconfigured notion of *debt* to breach the commonplace political gesture of seizing the proper that is found in all theories of community that are articulated through the dialectic of alienation and appropriation. I argue that each thinker turns to debt in order to think about the relationship between ontology, ethics, and politics in their writings on community. I shall conclude by revisiting Esposito's categorization of Italian Thought. Here I focus on the relationship between neutralization and conflict.[8]

The Dialectic of Alienation and Appropriation

In the first wave of the continental debate on community, the problem of seizure was largely configured within the broader problematic of economism, workerism, and productivity. From Nancy's *The Inoperative Community*, to Blanchot's *The Unavowable Community*, to Agamben's *The Coming Community*, a major reference was the refusal to seize power. Besides Bataille and Kojève, Agamben's notion of *inoperosità* was heavily influenced by operaismo and the many political philosophers this movement produced, just as Nancy's *désœuvrée* was influenced by Socialisme ou barbarie and the many political philosophers it produced. It was not until Nancy revisited his work in a series of papers in the 1990s, culminating in his magnum opus *Being Singular Plural*, and Esposito's response with *Communitas*, that the dimension of the proper was fully brought to the fore.[9]

In this debate, each author sought to reconfigure community in a manner that is not constituted through dialectic of alienation and appropriation. Put differently, each sought to conceive of community beyond the proper. The main authors do not give Pierre-Joseph Proudhon the attention he deserves. His early and crude ruminations on the "property prejudice" helped to establish a critical discourse on the role of property in mainstream political thinking.[10] We might even call Proudhon the progenitor of this critique. After over 160 years, what Proudhon viewed as a mere prejudice has metamorphized into the comprehensive, axiomatic, and hegemonic *dispositif* of the proper. In this *dispositif*, community is conceived as the finished product of the dialectic of alienation and appropriation, which gives rise to a tautological formulation of participation and sharing that has plagued all proprietary notions of community. Each member is expected to participate *in* the taking while simultaneously taking part *of* the participative activity. Each must appropriate the collective activity and render it proper to oneself; yet, precisely because it is appropriative activity that is appropriated by others, one is likewise appropriated by the activity. One shares in the activity, is shared out by the activity, and is thus supposed to be in a state of sharing with.

The basis for such a community is sharing. Without sharing, appropriation remains a private activity, which Esposito claims results in a "division without sharing" (*CM*, 28). The social contract, for example, is constituted as an agreement between private owners who relate with one another on the basis of self-interests. There is division here, but little sharing. However, when stretched to its logical ends, this formulation gives rise to a different problem: sharing without division. If there must be a proper, an ultimate appropriator, which we find in all the zero sum formulations of alienation and appropriation, individuals become appropriated by the collective and the collective becomes a hypostatic entity that absorbs individuals: sharing without division. To render relationships, each and everyone's, mine, yours, and ours, to combine the me and the us, is to negate the exteriority that is necessary to *be* in, not have, a relation with another. Relationships cannot remain relational if they are appropriated; put differently, proper relationships can only be improper. This is a core insight that has been developed in the debate about community. In fact, neither sharing without division or division without sharing are properly relational conditions. The proper, ultimately, is anti-relational.

Jargon aside, this strain in these texts is of paramount importance for our neoliberal era. Today, the political is not just defined within the horizon of the economic because the economic is simultaneously defined with the horizon of the political. The two have collapsed into each other such that it is nearly impossible to speak of one without enlisting categories originally belonging to the other. The traditional critique of political economy, that private interests dominate the public sphere, no longer carries the semantic breadth to adequately cover what has actually happened. It stands today as a naïve victim of its own accomplishments. Any solution it can proffer is immediately circumvented and recalibrated as yet another dimension of the *dispositif* of the proper. Regardless of where one is situated along the mainstream political spectrum, from left to right, when one speaks of politics he or she is forced to enlist economic categories. In our time, it is not just the public, or politics, that have been fused with the economic, but the political itself. And, to be clear, I am not claiming that the economic dominates the political, but that the political and the economic are now indiscernible. This problem can be traced back to John Locke's early formulation of the proprietary subject. In *The Second Treatise on Government*, he called for the liberal state to serve the "mutual *preservation* of the lives, liberties and estates, which I call by the general name, *property*"[11] or simply "*the preservation of their property*."[12] Agamben and Esposito have sought to rectify this predicament in their work, to which I will now turn.

Debt

In their texts on community, Agamben and Esposito are pressed to disrupt the reduction of *being* to *having*. Both draw from Heidegger's evental modality of ex-sistence in order to conceive of a notion of debt that allows them to reconfigure the relationship between ontology and ethics in an improper manner. This work becomes foundational for their later writings on biopolitics. In the following section, I primarily focus on their texts on community.

Agamben

> For me, the problem is to try to diagonalize [*diagonaliser*] the proper and the improper, to think, in logic as well as politics, of a language that is beyond the proper and the improper, this is what strikes me as the problem of belonging.[13]

Whether Agamben diagonalizes the proper and the improper or scribes a theory of the improper, as Timothy Campbell claims in *Improper Life*,[14] is a serious question that cannot be overlooked when reading *The Coming Community*. Agamben calls for an appropriation of the improper, of rendering the improper proper and thus one's own, but when translated into the practical terrain one wonders whether he diagonalizes or merely reaffirms impropriety and thus radicalizes difference. Bare life, for example, is the improper that is excluded and thus included within the semantic field of the proper. Whether he erases, neutralizes, or utilizes alienation and impropriety is a problem for his work, especially because improperness appears to be the very means and modality through which we are in common (communication). Being improper appears to be that which is our most proper.

In *The Coming Community*, Agamben argues that politics are obstructive of ethics. Politics are presuppositional and operative, whereas ethics are without presuppositions and inoperative. Politics create the abject conditions that render non-political, or post-political, solutions possible. This formulation has led many to question whether there is a politics to be found in Agamben, or if he merely prescribes an obscure and passive form of ethics that is for the most part impractical. He is often read as a philosopher who muses in messianic and eschatological texts, provides only cryptic prescriptions, and is, in the end, a negative theologian who treats the most abject conditions as a passageway into utopia. Because politics are presuppositional and operative, the only opening left for entering into a more ethical way of life is to conceive of liberation in an extremely passive manner. The two figures that epitomize this ethos in his book on community are the petty bourgeois and stateless people.

Despite their differences, Agamben and Esposito share a common goal of overcoming the privative notion of debt that is found in the contemporary articulation of the "economy of compensation" (*economia del risarcimento*). Both appeal to an alternative notion of debt by infusing this concept with an ontological and ethical inflection. Esposito's *munus*, for example, combines these dimensions in a deontological formulation that creates a communal sense of duties and obligations. Agamben, however, is not as eager to carry the ontological into the realm of relationships.

In *The Coming Community*, there are two main references in his theory of debt. First, debt is an ontological problem par excellence for Agamben. He draws from Heidegger's distinction between ontological and ontic readings of guilt in the passages on the "bad conscience" (*schelchtes Gewissen*) to make this clear.[15] Ontic guilt, for Heidegger, belongs to the model of compensation. To be guilty in this sense means that one *has* a debt, either because one owes something in return or because one has committed an act for which he or she is responsible. It refers to a "lack" or a "privation" that requires a measureable form of compensation. This is a type of debt, in other words, that can be appropriated and rendered whole again. Ontological guilt, on the other hand, is a guilt for which there is no solution, especially no compensation, because it concerns the fact that one *is*. "*Being-guilty*," Heidegger claims, "*does not first result from an indebtedness, but that, on the contrary, indebtedness becomes possible only 'on the basis' of a primordial Being-guilty*."[16] This is his famous thesis on our throwness in the world, of the impossibility of being responsible for our origin. All that can be accomplished in ontological guilt is to heed to the "appeal correctly," to understand "oneself in one's ownmost potentiality-for-Being—that is, to project[. . .] oneself upon one's *ownmost* authentic [or proper] potentiality for becoming guilty"—to accept "Being-the-basis of a nullity."[17] "Existing authentically," notes Raffoul in *The Origins of Responsibility*, "can only mean taking over or making oneself responsible for this 'not,' "one's finitude."[18] This sense of ontological debt also operates in Agamben's notion of the existential ethos:

> Since the being most proper to humankind is being one's own possibility or potentiality, then and only for this reason (that is, insofar as humankind's most proper being—being potential—is in a certain sense lacking, insofar as it can not-be, it is therefore devoid of foundation and humankind is not always already in possession of it), humans have and feel a debt. Humans, in their potentiality to be and to not-be, are, in other words, always already in debt; they always already have a bad conscience without have to commit any blameworthy act.[19]

The second reference in Agamben's notion of debt (*debito*) is etymological, which is a reference he shares with Esposito. The Latin term *debere*, breaks down as *de-habere* or *de-having*. It is here that Agamben hints at, but much less forcefully than Esposito's deontological prose, of the relationship between ethics and ontology by reversing the order of *being* and *having*. This is a form of debt that is beyond the order of appropriation because when one is in debt one is exposed and expropriated. This is a debt for which there is no compensation or remuneration because it is a type of debt that is, in Nancy's terms, incommensurable. One is rendered improper precisely because debt places one in a position that is "humankind's most proper being."

This double sense of debt opens up the possibility of rethinking the relationship between ethics and ontology. Ethics begins from the fact that there is nothing to "enact or realize" whether this be a destiny, essence, or history (*CC*, 43). Since he is critical of presuppositional thought and opposed to operative formulations, however, Agamben is careful not to appeal to any prescriptive or consequentialist model of ethics. Instead of an ethics proper, he turns to the *modality of being-thus*, which, he argues, is a particular *ethos* that disrupts presuppositions and operativeness. He plays with the connection between *ethos* and *habit* (both present in *de-habare*) to make this claim. The "only ethical experience," he contends, "is the experience of being (one's proper) potentiality, of existing (one's proper) possibility—exposing, that is, in every form of one's proper amorphousness and in each act one's proper in actuality" (*CC*, 44; translation modified).[20] Being-thus represents a particular modality of existence, or a "*manner of rising forth*," which is not a modality one enters into or takes up, rather it is "a being that is *its* mode of being" (*CC*, 29). Only in this manner, can we "find a common passage between ontology and ethics" (*CC*, 28).

Contrary to morality, a being that is rising forth cannot "*presuppose* itself" because it is *exposed* to its "qualifications" when it "*is* its *thus* without remainder" (*CC*, 28). The sheer force of the event liberates the self from the operative logic of the world. It is thus free to use itself however it sees fit because it exists in the liberated time of the brief sabbatical. This freed time is neither useless nor inactive, but a time where the activity of the *argos* does not make sense to the commonplace activity of the *ergon*. One is free to be creative, in the multiple senses implicated in this passage, which is why he claims in an almost prescriptive fashion that "the only happiness possible for humans" is to be "generated from one's proper manner" (*CC*, 29). As an existent one can appropriate his or her impropriety and treat it as if it is proper being. Yet, this being "does not belong to it" because "it is," he claims, "perfectly common" (*CC*, 29). Impropriety, he claims in a manner that is more contrarian than diagonal, "is our second, happier, nature" (*CC*, 29).

Ethics must adhere to this ethos. To be a proper host or neighbor, he argues, one must not compensate "for what the other lacks" (*CC*, 24). Ethical relations are only possible in the space of "ease" (*agio*), "the space adjacent (*ad-jacens, adjacentia*)," which is "the empty place where each can move freely" (*CC*, 25). It is the space of love. In this sense, ease "names perfectly" Hölderin's " 'free use of the proper.' " Evil, on the other hand, is found "in the decision to remain in deficit of existence," which is, of course, a modality where one seeks to "appropriate the power to not-be" and convert it into "a substance and a foundation" or to view potentiality "as a fault that must always be repressed" (*CC*, 44).

In all of his works, Agamben seeks to rethink the relationship between *ethos* and *modality*. This problem arose in his early writings on the pure means without ends (2000),[21] impotentiality (1999),[22] and inoperativeness (2007);[23] it was also articulated in his more applied biopolitical analyses of the camp (2002),[24] *homo sacer* (1998),[25] and the state of exception (2005);[26] and he has once again returned to it in his recent writings on economic theology (2011)[27] and destituent power (2014).[28] What he called being-thus gradually became translated into a biopolitical lexicon as "form-of-life." In his recent lecture on destituent power he makes this point clear. Form-of-life, he claims, is "a life that can never be separated from its form, a life in which it is never possible to isolate something like a bare life," which is "a life for which, in its way of living, what is at stake is living itself."[29] The stake for this life is precisely its "mode of living," the *ethos* in *The Coming Community*, where "the single ways, acts, and processes of living are never simply *facts*, but always and above all *possibilities* of life, always above all potentiality."[30]

In this lecture, Agamben also makes what might be his most succinct statements regarding his political project over the last two decades. It "will not be possible," he argues, "to think another dimension of life if we have not first managed to deactivate the *dispositif* of the exception of bare life."[31] Unlike Esposito, Agamben claims that it is not possible to directly access the *dispositif of the exception*, the inclusive exclusion of bare life.[32] If politics are to be inoperative and destituent, a philosopher can only (or should only) strive to exhibit and expose the form through which the *arche* of power operates. Such work was carried out in his *Homo Sacer* series. Bare life is a life that has been captured, excluded, and included as a *what*. As such, rather than being-thus, it is no longer a form-of-life. To seize, even to oppose and deconstitute, this *arche*, he argues, would give rise to a reconstitution of the *arche*. Thus, the sole political task available for Agamben is to exhibit, to lay bare, and expose the *dispositif* of the exception. Such a task can open up new possibilities for a truly modal ontology. This would be a modal ontology, an ontological ethos, where "*my form-of-life relates not to what I am, but to how I am what I am.*"[33] In the end, his modal ontology of de-having is a philosophy of exposure and de-positioning. Whether this results in a diagonalization of the proper is another question.

Esposito

> [T]he common is not characterized by what is proper but by what is improper, or even more drastically, by the other; by a voiding, be it partial or whole, of property into its negative; by removing what is properly one's own [*depropriazione*] that invests and decenters the proprietary subject, forcing him to take leave of himself, to alter himself. (*CM*, 7)

Despite their differences, there are many overlapping motifs and exigencies in Agamben and Esposito. Agamben, recalling last section, has recently called for a *deactivation* of "the *dispositif* of the exception of bare life."[34] Destituent politics, for example, do not seek to capture, appropriate, and reconstitute this *dispositif*; rather, they attempt to render it inoperative by exposing its operations. Esposito likewise emphasizes the opposition between *disponere* and *exponere* in his writing. In fact, this opposition is present in his core theoretical distinction between *immunitas* and *communitas*. The modern immunitary *dispositif* arranges, orders, controls, and regulates things, persons, and relationships in a negative and proprietary manner. It grants individuals a "pure negative right," he contends, "to exclude all others from using what is proper" to themselves (*IM*, 25). *Communitas*, on the other hand, is defined by its expository nature. In *communitas*, each is exposed, divided, and shared out (*condivisione*) such that immunized barriers, whether individual identity or national borders, are broken down and rendered inoperative. As such *communitas* can never be conceived as a thing or a place that is proper only to its members. It is thus inaugurated by an exposure that is so powerful that each is open to differences.

Esposito insists in all of his writings from *Communitas* onward, that community can only be realized through an interruption and transformation of immunity. Most of this work is carried out in his biopolitical writings. In a gloss of his work in the closing pages of *Living Thought*, he notes that the "burdensome law of the *munus*" provides a challenge to the "state of exception" found in modern individualistic immunization processes (*LT*, 258). Community cannot occur without an immunitary system. The point is to find a balance between community and immunity, for when "they diverge or exceed one another," he argues, "the consequences can be catastrophic" (*LT*, 261). The internal mechanisms of the immunitary *dispositif*, he continues, must "be shifted from a defensive barrier against the outside"—to remove all the "aggressive metaphors, or tropes" and "military jargon" from the biological lexicon of biopolitics—"to a process of differential transformation of the very subjects it identifies" (*LT*, 261). The point is to find more tolerant forms of immunization that are open to alteration, where immunity becomes a "filter for contact and communication with the surrounding environment" (*LT*, 261).

One of his most succinct statements on the political implications and exigencies of this problematic are found in his essay "Community, Immunity, Biopolitics."[35] He argues that the contemporary political "task at hand is to overturn in some way—indeed in every way—the balance of power between 'common' and 'immune.' " We must find ways to inaugurate the delicate procedure of separating the "immunitary protection of life from its destruction by means of the common; to conceptualize the function of immune systems in [a] different way, making them into relational filters between inside and outside instead of exclusionary barriers."[36] This task, he continues, must simultaneously negotiate between "two levels: by disabling [*disattivazione*] the apparatuses of negative immunization, and by enabling [*attivazione*] new spaces of the common." In his biopolitical writings, Esposito characterizes the *enabling* dimensions of his philosophy as affirmative. Affirmative biopolitics dis-activates the immunization effects of negative biopolitics while simultaneously enabling the affirmative dimensions of the *munus*. In this work, the ethical, political, and ontological dimensions of his philosophy are woven together. It is not my intention to cover his entire theory of *communitas* here; instead, I briefly elaborate on the deontological nature of the *munus*.

Esposito argues that the modern legal model of the private person has come to replace the impersonal subject of *communitas*. In *Living Thought*, he claims that the person "presupposes the separation from itself as opposed to the unity of the living being" (*LT*, 272–73). As with the notion of negative liberty, the person is drawn from a proprietary model of a subject who is the owner of their own body.[37] Under the regime of the person, he argues in *Immunitas*, law becomes a private law (*ius proprium*) that immunizes the community and "reverses the affirmative bond" of the *munus*. Common duties and obligations, he states in *Immunitas*, become "the purely negative right of each individual to exclude all others from using what is proper to him or her" (*IM*, 25). He has written extensively on the problems of the regime of rights and rights-based discourse. Rights, he argues in *Third Person*, derive from the "exclusionary *dispositif*" of immunity because a right is simultaneously "private and privative"

(*TP*, 101). Rights are thus always particularistic, even when applied to a category of people they necessarily exclude others.

Esposito often turns to Simone Weil to make the argument that we have to dis-activate the personal subject of rights in order to enable a notion of the *impersonal*, which opens up the possibility of thinking about communal obligations. The impersonal is anonymous. It disrupts the immunitary mechanisms of the law and opens unto justice. Weil, he argues in *Immunitas*, formulated a " 'law in common,' " which prioritizes "obligations" over "rights." Obligations are binding. Only in the law in common, are we subjected to "an expropriation of what is proper to us, beginning with the subjective essence" (*IM*, 23). This is, to return to my opening sentences of this section, an expository modality that disrupts the arrangement, ordering, controlling, and regulation of things in the immunitary *dispositif*. The affirmative element of these expository relationships is articulated in his theory of the *munus*. It is here that he weaves together the ethical, political, and ontological dimensions of his philosophy. I believe that this strain of his philosophy represents a radically deconstructed brand of republicanism, which one might call a continental rendition of communitarianism.[38] It is also in this strain where we find parallels with Agamben, especially in their efforts to disrupt the proprietary rendition of the economy of compensation with a reformulated notion of debt.

As discussed above, Agamben emphasizes the expropriating sense of *debere* (de-having). Ontological debt cannot be subjected to the logic of appropriation because it expropriates the subject. The connection between ethics and ontology in this formulation is not direct. Instead, to be in debt is to be a particular modality, habit, or ethos qua being-thus or, in his more recent writings, form-of-life, that acts as an ontological passageway that renders ethical relations possible. Debt is an ontological disposition that renders ethics possible. It is a peculiar opening without content. It is a modal ontology, not an ethical ontology.

Esposito, on the other hand, does not shy away from thinking the relationship between ethics and ontology. In addition to *debt*, he draws from another derivative of *debere*, duty, to draw a clear connection between ontology and ethics. In his ethical ontology, ethics is grounded in ontology and ontology in ethics. This is firmly established in his notion of the *munus*. Ontologically, *being* takes precedence over *having* in the *munus*. Ethically, the *munus* establishes a heterodox model of exchange. The *munus* opens up, transforms, and exchanges subjects; expropriates and diminishes them to the point that they are wholly lacking; and binds and indebts them to their contractual obligations (*CM*, 4). In the *munus* we are, in other words, contracted or drawn together, in "the transitive act of giving" that, recalling the economy of compensation, is not formulated according to the "stability of possession and even less the acquisitive dynamic of something earned, but loss, subtraction, transfer" (*CM*, 5). Here interest is reversed to its original sense, not a private interest, but the concern and importance of *being between* (*interresse*, *inter-esse*) (*CM*, 141). There is no sense of compensation or remuneration in this loss, as we find in more traditional social contracts, because this modality expropriates, exposes, and binds us together while simultaneously obliging us to perform services on behalf of the *com-munis*. Ontological debt gives rise to a deontological duty.

Furthermore, there is a direct relationship between his ethical ontology and politics. Politics are not left aside to the future (to come) or made possible in a particular ethos or way of being, instead, his affirmative politics are supplemented by the ethical duty implied by the *munus*. In his republican model, communal duties and obligations, which are ontologically grounded, are prioritized over rights and interests. There is a sort of impolitical opening here that allows for a rethinking of politics and the political.[39] We are drawn or contracted together (*con-trahere*) in sort of ontological contract that obliges us to give back to the gift of community. This is a heterodox deontological contract.[40] The politics that arise here are as much ontological and ethical as they are political. An affirmative biopolitics must affirm life, our ontological relationships, and the gift of community. Contrary to Agamben's modal ontology that emphasizes the expositional and preparatory characteristics of the event of existence, even Nancy's relational ontology, Esposito has sought to scribe an ethical and political ontology. This is neither a regional ontology, nor an ontology of the political, but a grounding of ethics and politics in the realm of first philosophy.

Esposito brings something new to this discussion by re-engaging with the modern Western tradition of political philosophy. Issues such as the validity of metaphysics and pluralizing ontological philosophy or notions of potentiality, communication, and productivity in the Western canon, serve as background materials that guide his examination of the Western tradition. From his most ontological work *Communitas*, to his subsequent biopolitical writings, Esposito's primary aim has been to traverse this tradition. He forces us to confront the place, or misplacement, of the common in political thought. How does the *dispositif* of the proper usurp the common, or, more directly, how does this *dispositif* empty the common of everything that is proper to it? It is precisely this treatment of the common as something that is proper that he interrogates. The proper has nothing in common with the common. The proper of the common, a phrasing our language forces us to employ, is neither proper nor improper. Modern political economy deceives us into thinking the common in terms of the proper. But, this is a discourse that at its core is privative. It is not suited to think the common, period. In its place, Esposito turns to an ethical economy, which in *Communitas* is more ontological but in his subsequent writings becomes more biopolitical. This ethical economy also addresses the core communist problem of redistribution, not as pieces of property (personal or otherwise) in the first instance, but in terms of an ethical ontology and later ethical-biopolitical notion of sharing (*condivisione*). Esposito radically deconstructs the giving-taking modality of the modern *dispositif* of the proper. His notion of the *munus* provides more solid grounds, a slightly more practical philosophy at least in relation to this debate, which makes his work stand out.

Conclusion

Two of Agamben's most prominent critics in Italy are Negri and Esposito. In *Bios*, Esposito argues that Agamben's reading of biopolitics is too negative. He has also noted in "Dialogue on the Philosophy to Come" that although Agamben's "politics of 'pure means' . . . is a suggestive formula," it remains "very indeterminate."[41] Negri has been less restrained. He has likened Agamben's notion of "bare life" to a " 'utopian escape.' "[42] In *Empire*, he and Hardt criticize the Bartlebian model of the refusal on the grounds that it merely indicates the "beginning of a liberatory politics." An "empty" and "solitary" refusal, they contend, "leads only to a kind of social suicide." What is really called for is a constitution of "a new mode of life and above all a new community."[43] Put in more traditional Marxist terms, the politics of refusal merely represent an initial phase of revolutionary action. At best, it represents the development of a private economic consciousness, not class-consciousness and certainly not a revolutionary consciousness that will give rise to revolutionary reappropriation. One might go even further and question the conditions of liberation he seeks. One might even ask how can the so-called "lumpen proletariats" of our time, those that are completely abandoned, hold any transferable political currency, especially when they are forbidden from using traditional political means for their cause? But this would be a step too far.

Agamben certainly does not prioritize conflict over neutralization—recalling the third pairing of Italian Thought in Esposito's recent paper.[44] At the same time, it would be unfair to say that Agamben only seeks to neutralize the *dispositif* of exception because although he does call for a neutralization of its operations and operative logic, he also calls for a destituent form of power.[45] The disruption of this *dispositif* likewise neutralizes the reduction of the political to the economic and the economic to the political that we find in the *dispositif* of the proper. Destituent power seeks not to appropriate that which has been deemed expropriated. It creates openings for rethinking politics and ethics. But, he stops short from making any prescriptions that dictate what is to be done and when he comes close to doing so he usually cuts himself off by inserting a messianic phrase, such as his call for a "politics to come." He also refuses to take the final step in his ontological philosophy by converting it into an ethical and/or political ontology. At most, his modal ontology renders possible a thinking about the relationship between politics and ontology or ethics and ontology. Whether his ontology of the how enables us to rethink politics in anything other than ecstatic or evental state, remains a problem for his philosophy.

Esposito is not unaware of the issues that have plagued or animated, depending on your preferences, Agamben's work. They belong to our times and to the tradition of twentieth- and now early twenty-first-century continental philosophy and politics. Like Agamben, he seeks to dis-activate the dialectic of alienation and appropriation, and the conflation of the economic with the political. However, in place of a modal ontology, or Nancy's relational ontology, Esposito scribes an ethical ontology that provides the contours for rethinking politics in a deontological manner. Politics are grounded in the duty/debt of the *munus*. First philosophy is, in other words, ethical and political or, in short, deontological.

Notes

1. In addition to Roberto Esposito whose hospitality was once again unconditional and Antonio Calcagno and Inna Viriasova for inviting me to participate in this collection, I must thank Wilfrid Laurier University, SSHRC, and the Scuola Normale Superiore di Pisa. I wrote this paper as a visiting researcher

at SNS. I would also like to thank the organizers of the *Convegno Internazionale Italian Theory* (Naples, Italy–May 2014)—Roberto Ciccarelli, Dario Gentili, Elettra Stimilli, and Davide Tarizzo—for inviting me to present a working version of this paper at their conference. This paper represents an updated, longer, and English version of the paper I presented at this conference, which was published in 2015 (Greg Bird, "Ripensare il proprio da Agamben a Esposito," in Dario Gentili and Elettra Stimilli, eds., *Differenze Italiane: politica e filosofia: mappe e sconfinamenti* (Roma: DeriveAprondi, 2015), 213–25).

2. In the English-speaking world, Lorenzo Chiesa and Alberto Toscano's edited volume *The Italian Difference* has helped popularize this discussion (Lorenzo Chiesa and Alberto Toscano, eds., *The Italian Difference: Between Nihilism and Biopolitics* (Melbourne: RE Press, 2009)).

3. Antonio Negri, *La differenza italiana* (Roma: Nottetempo, 2005).

4. See the debates published in the conference proceedings from the French and Italian Conferences in 2015 (Silvia Contarini and Davide Luglio, eds., *L'Italian theory existe-t-elle?* (Paris: Mimesis, 2015); Dario Gentili and Elettra Stimilli, eds., *Differenze Italiane: politica e filosofia: mappe e sconfinamenti* (Roma: DeriveAprondi, 2015); Dario Gentili, *Italian Theory: Dall'operaismo alla biopolitica* (Bologna: Mulino, 2012)).

5. Roberto Esposito, "German Philosophy, French Theory, Italian Thought," in Dario Gentili and Elettra Stimilli, eds., *Differenze Italiane: politica e filosofia: mappe e sconfinamenti* (Roma: DeriveAprondi, 2015), 15.

6. Ibid., 17.

7. Giorgio Agamben, *State of Exception*, trans. Kevin Attell (Chicago: University of Chicago Press, 2005); Giorgio Agamben, *The Kingdom and the Glory: For a Theological Genealogy of Economy and Government*, trans. Lorenzo Chiesa (Stanford, CA: Stanford University Press, 2011).

8. This chapter represents a précis of one of the lines of inquiry that I develop in my book *Containing Community*, which examines the continental debate about community. I have also drawn from a few key points I made in Greg Bird, "Roberto Esposito's Deontological Communal Contract," *Angelaki* 18, no. 3 (2013): 33–48.

9. Jean-Luc Nancy, *The Inoperative Community*, trans. Peter Connor (Minneapolis: University of Minnesota Press, 1991); Jean-Luc Nancy, *Being Singular Plural*, trans. Anne E. O'Bryne and Robert D. Richardson (Stanford, CA: Stanford University Press, 2000); Maurice Blanchot, *The Unavowable Community*, trans. Pierre Joris (Barrytown, NY: Station Hill, 1988); Giorgio Agamben, *The Coming Community*, trans. Michael Hardt (Minneapolis: University of Minnesota Press, 1993); Roberto Esposito, *Communitas: The Origin and Destiny of Community*, trans. Timothy C. Campbell (Stanford, CA: Stanford University Press, 2010). Of course, it was present in the earlier ruminations, especially in Agamben, but with Nancy's second text and Esposito's addition, it became clear that one of the main

dimensions of their joint efforts was to think community beyond the purview of the proper.

10. Pierre-Joseph Proudhon, *What is Property?*, trans. Donald R. Kelly and Bonnie G. Smith (New York, NY: Cambridge University Press, 2005).

11. John Locke, *Second Treatise on Government*, ed. C. B. Macpherson (Indianapolis, IN: Hackett Publishing Company, 1980), §123.

12. Ibid., §124.

13. My translation of a discussion Agamben had with Alain Badiou about *The Coming Community*. The original reads: "*Pour moi, le problème est d'essayer de diagonaliser le propre et l'impropre, de penser, aussi bien en logique qu'en politique, une langue qui soit au-delà du propre et de l'impropre, c'est pourquoi je me heurte au pb [problème] de l'appartenance*" (see Alain Badiou, "*Intervention dans le cadre du Collège international de philosophie sur le livre de Giorgio Agamben: 'La communauté qui vient,' théorie de la singularité quelconque,*" http://www.entretemps.asso.fr/Badiou/Agamben.htm (accessed January 29, 2012).

14. Timothy C. Campbell, *Improper Life: Technology and Biopolitics from Heidegger to Agamben* (Minneapolis: University of Minnesota Press, 2011).

15. Martin Heidegger, *Being and Time*, trans. John Macquarrie and Edward Robinson (New York: Harper & Row Publishers, Inc., 1962), §57–60.

16. Ibid., §58.

17. Ibid., §58.

18. François Raffoul, *The Origins of Responsibility* (Bloomington: Indiana University Press, 2010), 265.

19. Agamben, *The Coming Community*, 43–44. Hereafter cited in the text as *CC*.

20. "*[L]'unica esperienza etica . . . è di essere la (proprio) Potenza, di esistere la (proprio) possibilità; di esporre, cioè, in ogni forma la propria amorfia e in ogni atto la propria inattualità*" (Giorgio Agamben, *La comunità che viene* (Torino: Bollati Boringhieri, 2001), 40).

21. Giorgio Agamben, *Means Without End: Notes on Politics*, trans. Cesare Carino and Vincenzo Binetti (Minneapolis: University of Minnesota Press, 2000).

22. Giorgio Agamben, *Potentialities: Collected Essays in Philosophy*, trans. Daniel Heller-Roazen (Stanford, CA: Stanford University Press, 1999).

23. Giorgio Agamben, "The Work of Man," trans. Kevin Attell in Matthew Calarco and Steven DeCaroli, eds., *Giorgio Agamben: Sovereignty and Life* (Stanford, CA: Stanford University Press, 2007), 1–10.

24. Giorgio Agamben, *Remnants of Auschwitz: The Witness and the Archive*, trans. Daniel Heller-Roazen (New York: Zone Books, 2002).

25. Giorgio Agamben, *Homo Sacer: Sovereign Power and Bare Life*, trans. Daniel Heller-Roazen (Stanford, CA: Stanford University Press, 1998).

26. Agamben, *State of Exception*.

27. Agamben, *The Kingdom and the Glory*.

28. Giorgio Agamben, "What is a Destituent Power?," *Environment and Planning D: Society and Space* 32 (2014): 65–74.

29. Ibid., 73.

30. Ibid.

31. Ibid., 66.

32. Ibid., 72.

33. Ibid., 73.

34. Agamben, "What is a Destituent Power?" 66.

35. Roberto Esposito, "Community, Immunity, Biopolitics," trans. Zakiya Hanafi, *Angelaki* 18, no. 3 (2013): 83–90; Roberto Esposito, *Comunità, immunità, biopolitica*, http://www.benicomuni.unina.it/roberto-esposito-comunita-immunita-biopolitica.html, accessed August 1, 2014.

36. Ibid., 87–88.

37. See the introduction and a chapter called "Person, Human, Thing" in *Third Person* for his account of the proprietary basis of personhood in liberalism (*TP*, 1–19, 64–103).

38. Greg Bird and Jon Short, "Community, Immunity, and the Proper: An Introduction to the Political Theory of Roberto Esposito," *Angelaki* 18, no. 3 (2013): 1–12.

39. There is now a debate regarding the place of the impolitical in Esposito's thought in the English literature. For a positive account of the relation between his thinking of the impolitical, the *munus*, and the possibilities they present for affirmative biopolitics, see María del Rosario Acosta, "Hegel on *Communitas*: An Unexplored Relationship Between Hegel and Esposito," *Angelaki* 18, no. 3 (2013): 13–31. For a negative account see Bruno Bosteels, "Politics, Infrapolitics, and the Impolitical: Notes on the Thought of Roberto Esposito and Alberto Moreiras," *The New Centennial Review* 10, no. 2 (2010): 205–38.

40. Bird, "Roberto Esposito's Deontological Communal Contract."

41. Roberto Esposito and Jean-Luc Nancy, "Dialogue on the Philosophy to Come," trans. Timothy Campbell, *The Minnesota Review* 75 (2010): 84.

42. Cited in Carlo Salzani, "*Quodlibet*: Giorgio Agamben's Anti-Utopia," *Utopian Studies* 23, no. 1 (2012): 228.

43. Michael Hardt and Antonio Negri, *Empire* (Cambridge, MA: Harvard University Press, 2000), 204.

44. Esposito, "German Philosophy, French Theory, Italian Thought."

45. Agamben, "What is a Destituent Power?"

4

Against the Conspiracy. Revisiting Life's Vertigo

On Roberto Esposito's *Terza persona* and *Da fuori*

Alberto Moreiras

Beatitude Against Subjection

Within the time that marks the reconstruction of the ideology of the person after World War II, existence first, then language, then culture: these notions emerged, with varying degrees of hegemonic success, as fundamental regions of research and thought. There was existentialism, and then the linguistic turn, and later the cultural turn, and they inflected or constituted a history or a trajectory of reflection that, in retrospect, seems to have remained more or less, or for the most part, confined to the academic humanities. It is harder to discern whether the technopolitical turn, with its radical impact on the notion of the person, is first of all an event in the history of thought or whether it should be more properly considered a turn in the history of humanity. The increasing contemporary dominance of technoscientific protocols, even at the level of economic development, challenges and undermines most conceptions inherited from modernity at the most practical and experiential levels. The technosciences will make leaps within the span of one or two generations, and the political implications of promised developments in, for instance, neurophysiology, pharmaconomy, or genetics, not to mention informational sciences, will change even more drastically than they have so far many of the conditions of social interaction. Indeed, it is because we are already at the threshold of potentially revolutionary change in at least some societies that there is an urgency to the formation of critical and constructive thought on the various consequences of the technopolitical turn that seem to exceed the conditions under which a fixation on existence in the 1950s or on language in the 1960s and 1970s, or on culture in the 1980s and 1990s, was political.

Giorgio Agamben's 1996 essay "Absolute Immanence" was programmatic. Its importance does not reside in its focus on the technopolitical. Rather, Agamben turns to that which the technopolitical seems to threaten, which would constitute the substance of his thought from *The Coming Community* down to his latest work: what he calls "life," in the sense of "bare life," which is a notion found in the work of Walter Benjamin.[1] The confrontation between bare life and technopolitical deployment, or rather the fundamental attack on life as we have known it at the hands of technoscientific protocols and their political ramifications—biopower, in short—is to be submitted to a genealogical analysis on the basis of a new conceptuality. This is Agamben's 1996 program:

> The concept of "life," as the legacy of the thought of both [Michel] Foucault and [Gilles] Deleuze, must constitute the subject of the coming philosophy. First of all, it will be necessary to read Foucault's last thoughts on biopower, which seem so obscure, together with Deleuze's final reflections, which seem so serene, on "a life . . ." as absolute inmmanence and beatitude . . . for Foucault, the "different way of approaching the notion of life," and for Deleuze, a life that does not once again produce transcendence. We will thus have to discern the matrix of desubjectification itself in every principle that allows for the attribution of a subjectivity; we will have to see the element that marks subjection to biopower in the very paradigm of possible beatitude.[2]

Unpacking this program is a task to be pursued along two main poles: the play between subjectivity and desubjectification, and the play between beatitude and submission. In terms of goals, since a program for a coming philosophy must incorporate a teleology, it is perhaps easier to understand the importance of beatitude, given its conventional association with happiness, than it is to move toward a practice of desubjectification. If things are mired in obscurity, well then, the task of the coming philosophy is no doubt to contribute to their clarification. Agamben offers a number of thoughts on the way. I will sum up some of them, as they will be relevant for an adequate evaluation of the stakes in Roberto Esposito's *Third Person* (*Terza persona*).

Most of Agamben's essay is taken up with a celebratory analysis of Deleuze's last essay, "Immanence: A Life . . ." The notion of *homo tantum*, understood or expressed in Deleuze on the basis of a passage in Charles Dickens's *Our Mutual Friend*, seems crucial.[3] This refers, in Deleuze's formulation, to *a* life, that is, to a point when "the life of the individual has given way to an impersonal and yet singular life, which foregrounds a pure event that has been liberated from the accidents of internal and external life, that is, from the subjectivity and objectivity of what comes to pass . . . beatitude; or an ecceity, which is no longer an individuation, but a singularization, a life of pure immanence, neutral, beyond good and evil, since only the subject that incarnated it in the midst of things made it good or bad."[4] *Homo tantum* is *a* life, in the sense that it is that in life which remains free from the aporias of subjectification, and even from consciousness. The *homo tantum* is the locus of a "transcendental field" defined as a "pure plane of immanence."[5] For Agamben, with the notion of the pure plane of immanence Deleuze has taken an "irrevocable step beyond the tradition of consciousness in modern philosophy" whose radicality goes or would go beyond anything accomplished by Edmund Husserl, Martin Heidegger, Maurice Merleau-Ponty, Emmanuel Levinas, or Jacques Derrida.[6] In fact, the latter thinkers, Agamben says, to the extent that they are thinkers of immanence, will still want "to think transcendence within the immanent," so that "immanence itself [is] made to disgorge the transcendent everywhere."[7] But Deleuze's notion goes beyond his colleagues' "necessary illusions" toward the "most extreme thought" of "a principle of virtual indetermination, in which the vegetative and the animal, the inside and the outside, and even the organic and the inorganic, in passing through one another, cannot be told apart."[8] This is Deleuze's triumph: to the extent that "today, blessed life lies on the same terrain as the biological body of the West,"[9] as Agamben offers rather enigmatically, *homo tantum* will have given us something like an analytical lever to initiate a practical countermove to technopolitical subjectification. "A life" is "a matrix of infinite desubjectification" and it is therefore radical resistance to biopower.[10] In the notion of "a life" Agamben salutes Deleuze's identification "without residue" of desire and Being, and thus the principle of a new and totalizing vitalism commensurate to the present epoch.[11]

Deleuze's transcendental empiricism has discovered the plane of immanence on the basis of an *il y a* which is less than a whisper in the background: it is the irreducible, what obtains before both being and nothing, pure virtuality attributable to nothing.[12] Imagine a wound. The wound preexists you. A wound is later incarnated in a state of things, in lived experience, it is narrated, it is felt; it can even become a source of transcendence. But the wound, before action, is always already a singularity in the virtual field of immanence. For Deleuze "events or singularities impart to the plane their full virtuality, just as the plane of immanence gives virtual events their full reality. The event considered as non-actualized (indefinite) lacks nothing at all. It suffices to put the event in relation to its concomitants: a transcendental field, a plane of immanence, a life, a few singularities."[13]

We are now in a position to understand Agamben's program for a philosophy of the future a little better: on the basis of the fact that it is the modern understanding of subjectivity that has led us to where we are, we must watch over the possibility of a reintroduction of transcendental subjectivity, which is the ground of the possibility of the "necessary" illusions.[14] If a life is the matrix of infinite desubjectification, we must uphold the priority of *a* life over against any concrete life. There is a danger that beatitude, power, desire, even when theorized from a postsubjective position, become once again transcendental illusions. Beatitude and subjection lie on the same plane of immanence. The political battle is a matter of making the active side of power overcome the reactive side of power—the virtual against the actual.[15] We must play off beatitude against subjection. But is that enough? Or rather: is it not too much already? How does beatitude—how could it—take us beyond the aporias of a technopolitical administration of life?

The Zone of Indistinction

"The person, one could say, is that which, in the body, is more than the body" (*TPN*, 15)[16] which also means that the body is always less and other than the person. Is *homo tantum* commensurate to the task of a new destruction of the concept of the person? If it is, is destruction what we need? Esposito argues that whatever goes today under the concept of the person has enjoyed "an absolute ontotheological primacy" (*TPN*, 5) since Roman law, and that even today's jurisprudence makes of the person the very condition of possibility of the subject of rights (and vice versa) (*TPN*, 5).[17] The category of the person unifies man/woman and citizen, soul and body, law and life. And yet, today more than ever, and despite the category's ideological primacy, we witness a growing disparity between form and existence, subjectivity and the body, life and the law. This is Esposito's fundamental thesis on the critical side: the ongoing disjunction between life and contemporary conditions, that is, our basic biopolitical fracture, is not occurring *in spite* of the ongoing re-affirmation of the democratic—because universal—ideology of the person, but rather *because of* it.

It is a theoretical circle with a long history. Even if this history antecedes it, Roman law organized a powerful systematic separation between the person as an artificial entity and man/woman as natural beings to whom the status of person could be more or less accorded. This meant the presence of a zone of indistinction that made of many human beings something less than persons, that is, beings that, while human, were still somewhere between person and thing: the slave is only the extreme point of a social gradation, and perhaps not the most extreme. A partial reification of the human, even when it was a question of distinguishing the properly human from its internal fallenness into thingness or animality, has remained a constant in the long history of ontotheology.[18] Through Western time, and through different historical epochs down to our present, the human was differentially defined, to its glory or its doom, through its rapport with the thing that was inside it, and that always already made it susceptible to appropriation and domination even at the same time that it set the conditions of possibility for its freedom and autonomy.

Even if a person is precisely not a thing, there has never been a definition of the person outside the consideration of the essential and irreducible presence of the not-personal thing within the person, which always therefore contaminates the person. "At the bottom of this convergence we find the Aristotelian definition of man as rational animal . . . the biopolitical corporization of the person or the spiritual personalization of the body are inscribed in the same theoretical circle" (*TPN*, 16). In terms of our recent past, there is a difference in the systematic treatment of the differential nexus between the person and the body: for instance, the liberal conception, which assigns the property of the body to the person that occupies it, and the Nazi conception, which assigns it to state sovereignty. But, Esposito says, even this heterogeneity rests on a common trait: a productionist conception of life that makes the body into a property, a piece of property. Contemporary bioethics still reproduces the Roman separation between *persona* and *homo* when theorizing the differences between the full person, the semi-person, the non-person, and the anti-person (*TPN*, 17), which is ultimately the theorization of the technopolitical conditions for the power of decision over life and death, and for everything in between. In the face of technoscientific development we need alternative understandings of the human, and even alternative understandings of the very notion of life.

And yet, Esposito says, "the logic of the person does not occupy the entire contemporary horizon" (*TPN*, 18). There is a thought of the impersonal that is beyond philosophy and can also be traced in literature and painting, in music and film, even in science. "The impersonal . . . is the mobile border or the critical margin that separates the semantics of the person from its natural effect of separation" (*TPN*, 19). The thought of the impersonal is not a philosophy of the anti-person—it only dwells elsewhere. The notion of a "third person" was suggested by Émile Benveniste's work on personal pronouns.[19] For Benveniste, the "third person," precisely, does not have personal connotations, but refers to another register that is able to bypass dialogical subjectivity and reach the region of singularity, even of plural singularity. To explore this region, to interrogate the possibility that the third person, or the impersonal, can undo the ontotheological primacy of the person (in ways other than thanatopolitical) and can thus provide for the possibility of a thought outside the theoretical circle of productivist life, is the final object of *Third Person*. My purpose in what follows is to focus on some analyses that seem particularly relevant to its understanding, and which I will link to references to Giorgio Agamben's *The Open*.[20]

The Shadow of the Impersonal

How does one make a third-person politics? The third person is not just another person, but rather it opens a different regime of sense. For Esposito, like for Agamben, Foucault and Deleuze are also key figures in the philosophical tradition that has moved against dialogical models in favor of a new theorization of the neuter—the neuter is not "some other that adds itself to the first two, but rather that which is neither the one nor the other, that which escapes every dichotomy founded, or presupposed, by the language of the person" (*TPN*, 21). For Esposito, Foucault and Deleuze will have succeeded in turning the impersonal not just into a force able to destroy (or "deconstruct") the *dispositif* of the person but, even more importantly, into "the form, or rather the content, of a practice that modifies existence" (*TPN*, 22). And Esposito calls "life" the figure of the third person in Foucault and Deleuze. (It should be noted that Esposito links here the new thought of life to what he will later call "neutralization"—and this is one of the sites from which Esposito will revise and then develop his own thought. More on this ahead.)

"Life . . . is for Foucault that biological side that never coincides with subjectivity because it is always caught up in a process of subjection and subjectivation—the space that power invests without ever totally occupying it, thus generating always new forms of resistance" (*TPN*, 23). If Foucault's notion of resistance—of life's resistance to subjection/subjectivation in the relentless push toward the "outside"—can become the "uncertain" basis of an "affirmative biopolitics" against contemporary power/knowledge (but not, of course, against science as such), then Esposito thinks of the third person as the site for a new political practice against threatening technopolitics, that is, for an affirmative and postsubjective technopolitics. And Deleuze's notion of "animal becoming" "opens the thought of the impersonal" not just to the "liberation from the fundamental prohibition of our time" but also to "the always promised but never truly experienced reunification of form and force, mode and substance, *bíos* and *zoé*" (*TPN*, 24), that is, to a new univocity of being where, like in Agamben, desire and being, or being and thinking, would coincide without residue. Esposito therefore also situates himself in the legacy of Deleuze and Foucault in terms of a coming philosophy now explicitly defined as a modification of existence, a practical politics of the third person (or a practical politics of "neutralization") that is also an affirmative politics of the impersonal. What is at stake is the rupture of the theoretical circle of the person and its originary separation or split of the human between rational and animal, person and thing, subject and object, spirit and body, substance and mode, even thinking and being. Foucault and Deleuze are not Esposito's only companions in this endeavor of beatitude.

But whatever may remain uncertain in the search for a positive biopolitics must be read in the context of the genealogical analysis. Esposito's first chapter posits a first beginning for it in the "paradigmatic turn" that came to affect or infect nineteenth-century political theology at the hands of biology, and in particular through the work of Xavier Bichat. Bichat's two main contributions for our purposes can be summed up in the polemological understanding of life as the set of functions resisting death ("the measure of life is . . . in general the difference that exists between the effort of the external powers and internal resistance. The excess of the former is life's weakness; the predominance of the latter is the index of its strength" [Bichat quoted in *TPN*, 26]), and the notion of two lives or a double life, split into organic life (vegetative functions) and animal life (which regulates relations with the outside). Death is the enemy, and its last triumph is its taking over of organic life. As Esposito shows, Bichat's work will have had tremendous impact on the understanding of the relations between the living subject and the conditions for political practice over the last two centuries, contributing to the rise of bio- and technopolitics against the founding notions of modern political theology, including modern democracy. In fact, the very presuppositions of the latter, such as the social pact between subjects endowed with rational will, and able to determine their own actions, will have suffered at least temporary collapse at the ideological level precisely through the consequent reduction in the notion of the person.

> Divided into an 'inside life' and an 'outside life,' into vegetative and organic life, life is traversed by an alien power that determines instincts, emotions, desires in ways not reconducible to a unique element. It is as if a non-man—something different and prior to animal nature itself—came to be inserted into man . . . From that moment the function of politics—now inevitably biopolitics—will not be to define relationships between men but rather to individuate the precise point of the border between that which is human and that which, inside the human itself, is other than human. (*TPN*, 30)

Biopolitics in its modern formulation is thus, to begin with, the decline of the person, and of the ontotheological primacy of the person. Arthur Schopenhauer and Auguste Comte are among the first thinkers who take Bichat's position seriously in terms of its implications for an understanding of the political. There can be no passage from the state of nature to the civil state when men are governed and controlled by their organic life. The function of the State is no longer to produce freedom, but rather to transfer the empire of death from the inside to the outside. Comte's "biocracy," by substituting *bíos* for *demos*, places at the center of politics the life of an organism, whether individual or collective, that must no longer abide by representation. Self-preservation is now all. This is far from Kantian republicanism.

Anthropology will be the scientific practice charged with drawing the consequences, and also linguistics, and then racial discourse. The power of life must be strengthened following laws of organic evolution that no longer respond to the categorical imperative and follow instead an entirely other logic. The human sciences have transferred their object from history to nature (*TPN*, 45), which will partially lead biopolitics in the direction of a thanatopolitics defined by the absolute superiority of the racial datum. The humanization of the superior animal is only the other side of the animalization of the inferior man (*TPN*, 65). The relative opposition between *bíos* and *zoé* finds its maximum stretch in the particular humanism of Nazi antropozoology: "Never like in this case . . . *bíos* and *zoé*, form of life and life without a form, are separated into a distance that is irremediable because it is constituted by the direct or inverted relationship with death: on the one hand a life so alive that it can propose itself as immortal, on the other hand a life no longer such—'existence without life,' *Dasein ohne Leben* . . .—because from the beginning contaminated and perverted by death" (*TPN*, 71). With it, the human, reduced to its position in a legacy of blood, race, species, became only that which remained from the destruction of the person that the French Revolution had made tendentially coextensive with the citizen. The mask had been abolished (*TPN*, 73).

The 1948 Universal Declaration of Human Rights, from which, arguably, our contemporary understanding of liberal democracy emanates, attempted to reconstruct it. If the emphasis of the 1789 Declaration had been on the notion of citizenship, now, in the wake of the radical depersonalization Nazi thanatopolitics had imposed, the dignity and value of the human person as such is vindicated and put at the center of the juridical and philosophical reconstruction of democratic culture after World War II. But with it, unavoidably or irremediably, the old dualisms—mask and face, image and substance, fiction and reality (*TPN*, 92)—came back to life, and to the notion of life. One consequence of making the person the subject of rights—of human rights—is to restitute the separation that keeps rights away from the non-person. But the non-person is always the other side of the person, and its fundamental condition. There is, in other words, no person without the non-person, as no mask can be recognized as such that does not imply the counterpresence of a face, as no appearance can exist without a founding presence, and as no fiction exists without the counterpositing of a reality, albeit merely imaginary, in the background. Already Hannah Arendt, in her book on totalitarianism, had recognized one key aspect of the aporetic structure that threatens the normative validity of the human rights of the person with the abyss: "the law admits only those who enter it from a given category—citizens, subjects, even slaves, to the extent that they are part of the political community" (*TPN*, 87). But those who are excluded from the categorial structure—from the structure of belonging, at any level—can only enter into law by breaking the law. The separation between the merely human and the person as a subject of rights still radically obtains. Those two elements, Esposito says, only meet "in the form of their separation" (*TPN*, 91), as the contemporary refugee crisis in Europe makes once again all too painfully clear.

Esposito's brief analysis of the Hobbesian inversion of objective rights into a subjectivistic notion of law is determinant. In principle, once the individual, beyond any difference in social status, is considered the bearer of a rational will and the total master of itself and its property, the Roman differentiation between man and person seems to lose its very foundation. The French Revolution, and its principle of equality, would have buried the Roman superordination of the subject to the law. But it is precisely at that moment that a new internal differentiation makes its appearance that reproduces at another level the gap between the individual as political subject and its undifferentiated humanity. Hobbes's notion of "artificial person," understood as the person of the representative, or the sovereign, already marks the dimensions of this irreducible gap in the political structure of modernity. The sovereign is "the only agent of personalization to the extent that before its institution no one can in the proper sense define himself as a person, either natural or artificial, because in the state of nature everybody coincides with their own living, and hence dying, being—that is, the transcendence that constituted the necessary condition of personality does not yet exist" (*TPN*, 104–05). But from the very moment that the sovereign emerges as an agent of personalization his role in universal depersonalization also emerges. The social pact, which constitutes the agent of the pact as a person, is simultaneously its depersonalization. Sovereignty marks the body of the "natural" person by "turning the person into that which no longer has a body and the body into that which can no longer be a person" (*TPN*, 107).

Adapting Sigmund Freud's definition of melancholy, one could say that, from the first modernity, the shadow of the non-person has always already fallen upon the person. The *dispositif* of the person traces indeed a melancholy paradigm, as the person can never fully emerge from its underside, which constitutes its permanent state of exception. The infinite effects this phenomenon has upon the contemporary administration of life remain to be worked out in critical analysis. Nazism is no longer our direct enemy, in the sense that it does not directly threaten us. Rather, liberal death spreads its wings today, still in the wake of Roman law but immeasurably enabled by contemporary technoscience, through its claim to organic life and its increasing encroachment upon human relationality, which includes the human's relation to animals and to its own body understood now, through a not-so-progressive turn of the screw, as what one is rather than what one has (present being now erases history, and a life's career cannot compete with present decline for any number of baby-boomers when it comes to labor relations, for instance). Esposito refers to issues of biotechnology, intangible property, and patents concerning vegetal and animal life as contemporary sites of deployment of the radical structure of separation at the core of the living from the perspective of the *dispositif* of the person, which has proven incapable of arresting the "progressive reification of life" (*TPN*, 117). Contemporary bioethics (Peter Singer and Hugo Engelhardt in particular come in for harsh criticism) does not hesitate to insist upon the animalization of liminar zones of the human under the dual thesis that "not every human being can aspire to the qualification of person, and not all persons are human beings" (*TPN*, 119). But of course these are only punctual examples of a political macroprocess of disenfranchisement that the contemporary ideology of the person not only does not prevent, but fosters.[21] In yet another paradoxical turn, it is precisely the pretension that the person is master of itself and its property, but only insofar as the person *is* what it has always been, that leaves not just so many of our fellow humans outside their own categorical enfranchisement (the non-persons, the quasi-persons, the semi-persons, the no-longer-persons, and the anti-persons Singer theorizes, for instance) but that constantly threatens to destroy even those of us who are otherwise secure or, in our best moments, believe ourselves to be secure in our present condition as persons (think of the current emphasis on post-tenure review in contemporary U.S. universities, for instance). Esposito's main (critical) thesis, namely, that the contemporary predicament has not developed in spite of the overwhelming ideological dominance of the *dispositif* of the person but rather because of it, has a crucial consequence: the need for the elaboration of an alternative understanding, without which no ideological struggle can be effective. The theoretical circle must be broken in order to give way to Esposito's positive thesis.

Against Subjectivation

The thinker Esposito invokes as the first radical proponent of a philosophy of the impersonal is of course Simone Weil, whose work in the 1930s already rose against the personalist ideology that many segments of the European liberal (and Catholic) intelligentsia were proposing as an alternative to fascism. For Weil the person depends on the collective, and right depends on might. From this perspective, for Weil both the category of the person and the notion of rights are complementary factors in what Esposito calls an "immunitary drift" whose end is the protection of privilege against the excluded.[22] Weil looks at the notion of person from the perspective of what it excludes, even as she also looks at rights from the perspective of what they steal. In other words, right is designed to protect the person against the non-person, which is always the non-person that has been defined as such from the very perspective of the right: this is the immunitarian drift. There can be no "universal right" of the person, since right is the mark of a communitarian privilege which is always had against the community's outside. The category of the person is for Weil, accordingly, a category of subordination and separation that must be fought through a radical appeal to the impersonal. "What is sacred, far from being the person, is that which, in a human being, is impersonal. All that is impersonal in the human is sacred, and only that" (Weil quoted in *TPN*, 124).

The passage to the impersonal: this is Weil's political demand. It is a passage beyond the I and the we, and therefore a passage into the third person, into the nameless or anonymous. The radically republican question is indeed of a pronominal nature. Is political justice, and political freedom, to be accomplished through the constitution of a we, or through the passage to the impersonal they? If my freedom is the freedom of all, is all to be encompassed by a first-person plural or by a third-person plural? Is political freedom a question of community or is political freedom a question of the multitude? Esposito will conflate the two parts of that question into one at the end of *Persons and Things*, and then again in *Da fuori* (*From Outside*).

Right around the time that Weil was dealing with these ideas she spent a few months in civil-war Aragon, close to the front. Had she been able to look beyond the trenches, into the other or anti-Republican side, she might have seen a few women with Y's patched onto their blue shirts. They would have been members of the Sección femenina, the Spanish Falangist organization for women, created and developed by Pilar Primo de Rivera. In Paul Preston's words:

> The symbol of the *Sección Femenina* was the letter Y, and its principal decoration was a medal forged in the form of a Y, in gold, silver, or red enamel according to the degree of heroism or sacrifice being rewarded. The Y was the first letter of the name of Isabel of Castille, as written in the fifteenth century, and also the first letter of the word *yugo* (yoke) which was part of the Falangist emblem of the yoke and arrows. With specific connotations of a glorious imperial past and more generalized ones of servitude, as well as of unity, it was a significant choice of symbol.[23]

So you are a woman, but you have subjectivized yourself as a person in an affirmation of love to the *Falange*. Your choice for the *Falange* is your personal freedom, but that freedom is, first of all, imperial freedom, as it commits you to a path of domination of others, the non-Falangists; secondarily, it is also imperial freedom to the extent that you sign up for your own domination, for your own servitude. You choose a collectivity that will not take its eyes away from you. As a member of the Sección femenina, it was your duty to serve the man, the men of the Fatherland, those fascists that you loved. Is Pilar Primo de Rivera and, with her, all the colleagues who thought up the Y symbol to sum up the free presence of Spanish women in the National Movement giving us the conditions of possibility of all political subjectivation? How does one become a person, politically speaking?

The community of the we is always the Y on your shoulder. The passage to the impersonal is the refusal of the Y. The uncanny choice for the freedom of all, for the freedom of the third-person plural, is a choice to be made outside and even against political subjectivation. It is adrift, as it refuses every orientation beyond itself, beyond its own gesture. It embodies no calculation, no teleology, no program. It is rare—rarer than the emergence of the subject itself, which happens every time there is a free choice for community. It stands outside every moralism (as it never seeks personal advantage). It is time to return the impersonal to the heart of the political. Everywhere we hear definitions of politics that presuppose political subjectivation as the goal. There is no doubt that political subjectivation is ongoing in every political process. But political subjectivation is in every case a function of the history of domination. The passage to the impersonal is the attempt to produce politics as the countercommunitarian history of the neuter.

Countercommunity

The significance of Benveniste's essay on the third person is high, to the extent that Benveniste, in Esposito's interpretation, has given us the *grammatical* conditions of possibility for the development of a sustained thought of the non-subject as the (logically) only possible thought of alterity: "Notwithstanding all the rhetoric about the other's excess, in the confrontation between two terms, [alterity] can be conceivable only and always in relationship to the I—its other side and its shadow" (*TPN*, 129).[24] If the I, confronting it, depersonalizes the you, it only does so to the extent that it awaits its own depersonalization in the reversal of the positions: the you always responds. The third person breaks away from the relationship between a "subjective person" and a "non-subjective person" by creating the possibility of a non-person: "The 'third person' is not a person; it is rather the verbal form that has the function of expressing the non-person" (Benveniste quoted in *TPN*, 131). The third person, beyond the I and the you, always refers to the absence of the subject, even if it can simultaneously refer to potential subjects. It is constitutively impersonal, and it is because of it that it can have a plural: "Only the 'third person,' as non-person, admits a true plural" (Benveniste quoted in *TPN*, 132).

It is from this position that a reading of Levinas opens up for Esposito, not as a thinker of the third person, but rather as the thinker who could not bring himself to the exposition of his own radicality. We are used to thinking of the Levinasian face-to-face as the epitome of Levinas's philosophical or antiphilosophical position. What Esposito's reading brings out, accurately, is the fundamental impossibility of the second-person suture in Levinasian thought—something that Levinas himself recognized, of course, and at the same time left undeveloped. At the limit, it would be something from which Levinas would have recoiled.

Esposito says that the question of the third person is for Levinas both "the theoretical vortex and the point of internal crisis" of his thought (*TPN*, 146). But, far from neutralizing it into the I-you encounter, Levinas recognizes the very originarity of the face as the trace of a field of signification that breaks every binary relationship: "the beyond from which the face comes is the third person" (Levinas quoted in *TPN*, 146). The beyond, which for Levinas means beyond being, also therefore means beyond transcendence, or beside transcendence. But this is the key problem. In the recognition of the third person as the beyond of the face, Levinas's thought opens itself to an unthinkability whose key position at the limit of twentieth-century thought makes it all the more urgent for us.

It is the problem, not of the impersonal, but of the impersonal's political import: the point or limit at which politics should no longer be thought of as contained in a dialogical structure is also the point at which politics abandons its all-too-human face in favor of a dimension able to affect, beyond and beside the third person and its infinite plurality, what Blanchot came to call the neuter for lack of a better term. Just like language "is spoken where a community between the terms of the relationship is missing" (Levinas quoted in *TPN*, 147), a countercommunitarian politics is a politics no longer structured in terms of friendship or enmity, no longer structured in terms of the interhuman relation.

Biopolitics finds its limit in the fallen dialogics of the subject/object relationship: it is the tendential application of technique to life (it is hence a technopolitics, but not the only possible one), for the purposes of an administration of life where life occupies the place of the object. Biopolitical practice, always modeled on the person's *dispositif*, is a practice of the master subject over against an object that constitutes it, and that by constituting it occupies the position of internal interlocutor. If the purpose of biopolitics is to make life sing to the tune of the subject, then it should be clear that no positive or affirmative biopolitics—all biopolitics is affirmative, even the Nazi kind: thanatopolitics is but the dark side of an essential affirmation—will suffice. A radical politics of the third person, hence beyond or beside the person, hence anti-biopolitical, finds its point of departure in Levinas's problem, his theoretical vortex and his point of crisis, which we can here only gloss following Esposito's indications, with a twist.

If the other is to command radical priority, there can be no common ground between the I and the you—the face comes up from a region of radical separation, or the you would become just another aspect of the

I. The other is not just a fold in a communitarian continuum, but the signal mark of an essential lack of community, and therefore the opening of and to a radical dissymmetry. If the subject suffers expropriation in Levinasian thought, it is because the demand of the other presses upon it from a region incommensurate to community. The experience of the you, when the you is not to be handled according to everyday linguistic convention but comes to us in the form of the face, radically, is then precisely at the same time the experience of that which can never be reduced to a you: an experience of the "third person," or of what Levinas calls "illeity." Sensing the beyond of the face of the other is at the same time encountering the third person.

But the third person recedes, and only seems to come in the form of its absence. It marks, in the first Levinas, a negative experience that might be referred to God as the Unreachable. But Levinas will later say: "Proximity is troubled and becomes a problem with the entry of the third" (Levinas quoted in *TPN*, 149). The third is a problem: recession breaks proximity, and proximity can no longer suffice. What is often ignored by Levinas's critics is that troubled proximity, and not the ethical relation, is the site of politics, which means that politics is the region that opens up in and through the very impossibility of community, in the rupture of the immediate ethical relation. It is through the very tension between proximity and its rupture (which is also at the very same time the rapture of proximity), or through the resistance to that tension (as the subject remains hostage to the other), that justice appears as the horizon of the political in the wake of the failure of the ethical relation to constitute itself as closed or unique horizon. This is what organizes the political as an insurmountable contradiction between the infinite ethical responsibility for the unique other, which introduces a radical limitation in the universality of law, and the equally infinite demand for justice, which is a limitation of ethical responsibility. Politics is for Levinas, to start with, this unstable field of relation created in virtue of the theoretical vortex that makes justice, as a demand that originates in the troubles of proximity, and ethics, as a demand imposed by the face of the other, equally unconditional. Esposito calls it a conflict between "partiality and equality," which, he says, reverses "the language of the person . . . into the form of the impersonal" (*TPN*, 152). The entry of the impersonal remains a problem because, with it, the subject is liquidated: not even as a hostage can it remain the source of agency. And this is something about which Levinas left few, and at best ambiguous, indications. The question, however, arises as to what the best way of supplementing Levinas's ambiguity would be. What if Levinas had been right in affirming the impossibility of a communitarian politics?

Esposito claims that it was Blanchot who made it his business to develop the Levinasian point of crisis into the insight of the neuter, "against the hostility, or at least the incomprehension, of the entire philosophical tradition" (*TPN*, 156–57). Blanchot mentions a "relationship of the third kind" which is precisely the disaster of every dialectics, of every dialogics, as the relationship that interrupts reciprocity and that therefore opens the non-relationship. The neuter is a non-personal alterity for which Blanchot rejects the name of "impersonal" as still insufficient (since "impersonal" is grammatically still dependent on a notion of person). Blanchot is looking for a break of the semantic field that will not allow it to reconfigure itself around the usual categories: being and nothing, presence and absence, internal and external. The third kind is the kind that enters no kind: and the neuter a word too much, which Esposito will link to the Levinasian notion of the *il y a* as it was developed in *De l'évasion* and *De l'existence a l'existant*. But for Blanchot the neuter is not primarily a site of existential horror; as the inevitable and destined site of existence, it is rather the "extreme possibility" of thought (*TPN*, 159).

What would be its political manifestation? A politics of the neuter is a politics of the third person in the sense already specified: an impersonal politics of the singular plural, a countercommunitarian politics of the they. Esposito's contention is that only Foucault and Deleuze were able to advance Blanchot's project. This is something that Giorgio Agamben has also sustained. As explained above, for Agamben the active category in the program for a philosophy of the future is the category of "life." Esposito connects the development of the category of life in the later thought of Foucault and Deleuze to a basic Nietzscheanism in both thinkers—to their emphasis on the notion of "force," which will be linked to an irreducible and untamable outside that is, however, and in virtue of its radical univocity, also our most intimate inside. "What is it that we are—beyond or before our persons—without ever taking possession of? What crosses and works us to the point of turning itself inside out if not life itself?" (*TPN*, 168). Power, also constituted by life, and to the extent that it turns itself against the human, never has enough with the person as subject of rights, but must go beyond the person and its end, beyond death therefore, toward the capture of life itself. Life captures itself as power, but at the same time life exceeds itself as force beyond power. This is for Esposito the very possibility of an "affirmative biopolitics" (*TPN*,

170) that he identifies with a new possibility of community, beyond the person, "singular and impersonal" (*TPN*, 171). "Life itself . . . constitutes the term on which the totality of the theory of the impersonal seems to be summed up and projected towards a still undetermined configuration, but because of that loaded with unexpressed potentiality" (*TPN*, 179).

A politics of the neuter is expressed in Esposito through his notion of a politics of impersonal life, even a biopolitics of impersonal life, that must lead, through the tapping of its unexpressed potentiality, not just beyond "the entire conceptual apparatus of modern political philosophy" (*TPN*, 179), but toward a new community, impersonal and singular: a community of beatitude which is, finally, the beatitude of the animal, the goal of the Deleuzian "animal belonging" that receives full recognition in the last pages of Esposito's book. While fully endorsing Esposito's deconstructive analysis of the person's *dispositif*, I have already expressed my objection in the form of a reserve regarding the possibility of an "affirmative biopolitics" that would finally render the metaphysical separation between *homo* and *persona*, which also means, between person and animal, null. It is ultimately an objection against beatitude. Agamben's *The Open* unquestioningly shares many of Esposito's insights and advances the argument toward a perhaps more nuanced understanding of the political task at least at the theoretical level.

The Critical Threshold

On the face of it, Agamben's *The Open* seeks a reconciliation of "man with his animal nature."[25] Its departure point was also treated by Esposito: it is Kojève's intriguing reflections about "the figure that man and nature would assume in a posthistorical world, when the patient process of work and negation by means of which the animal of the species *Homo sapiens* had become human reached completion."[26] If the destruction of the animality of man could be taken to sum up the humanist project down to the Hegelian system, for Agamben, glossing or critiquing Kojève, "perhaps the body of the anthropophorous animal (the body of the slave) is the unresolved remnant that idealism leaves as an inheritance to thought, and the aporias of the philosophy of our time coincide with the aporias of this body that is irreducibly drawn and divided between animality and humanity."[27] The body of the slave—*homo tantum*. We could define *The Open*'s project as the problem of reconciling *homo tantum* with the "physiology of blessed life" that a risen body—the posthistorical body—would want to enact. It is therefore a problem of beatitude, which remains, explicitly in Agamben, the eschatological horizon, and thus the orientation for an impersonal politics of the *whatever*.[28] We are back into a "critical threshold," and it is not unrelated to the Levinasian one. If for Levinas the threshold was defined by the incompatible unconditionality of the third person's demand for justice or equality and the particular other's demand for radical priority, for partiality, the threshold in Agamben is also a political threshold, that is, a threshold that determines a politics. It is the troubled separation between animal and human: also a trouble in proximity, seemingly irreducible for Western metaphysics, but which the technopolitical turn may be about to handle dismissively and with few compunctions.[29]

Agamben defines humanism's "anthropological machine," on the basis of Pico, Linnaeus, and others, as an "ironic apparatus" that must treat human nature as precisely an absence.[30] The human, caught in an endless play of specular misrecognition and projection, ends up establishing a "zone of indifference," a "space of exception" that is "in truth, perfectly empty, and the truly human being who should occur there is only the place of a ceaselessly updated decision in which the caesurae and their rearticulation are always dislocated and displaced anew. What would thus be obtained, however, is neither an animal life nor a human life, but only a life that is separated and excluded from itself—only a *bare life*."[31] The object of a political practice commensurate to the present, which on its positive side would be to promote beatitude, is, on its negative or critical side, to destroy the space of exception, that is, to dismantle the apparatus for the production of bare life. This is still a Foucaultian or Deleuzian politics of life, hence ultimately Nietzschean, which requires the division of force into active and reactive elements for the sake of the former. The active force of beatitude confronts the reactive force of error, not as truth confronts falsity, since truth could only emerge as the other side of error and hence still on the side of the reactive force, but rather as something deeper or taller: as an eschatology, Benjaminean, in any case un-Nietzschean. Early on in the book Agamben says: "perhaps even the most luminous sphere of our relations with the divine depends, in some way, on that darker one which separates us from the animal."[32]

The second half of *The Open* is almost entirely devoted to a partial analysis of Martin Heidegger's 1929–30 seminar *The Fundamental Concepts of Metaphysics: World, Finitude, Solitude*. In the 1996 essay discussed earlier, Agamben had placed Heidegger at the center of a genealogy of thought that included Immanuel Kant and Husserl on the one hand (the side of transcendence) and Baruch Spinoza and Friedrich Nietzsche on the other (for immanence), and that would result in Levinas and Derrida over against Deleuze and Foucault on the post-Heideggerian side: Heidegger's concepts are "the first figures of the new postconscious and postsubjective, impersonal and non-individual transcendental field that Deleuze's thought leaves as a legacy to his century."[33] Deleuze, counterintuitively, would have completed Heidegger's thought. The return to Heidegger now retraces the emergence of a political philosophy of the impersonal. But is it certain that the path marked by Foucault and Deleuze on the side of immanence must leave aside Levinas and Derrida as epigones of ontotheological thought that are therefore complicitous with the establishment or continuation of the apparatus of bare life? If Agamben's genealogy is to work, and obviously Esposito gives it some credence, we would have to consider Levinas and Derrida as thinkers of life's power against life, as reactive thinkers unable to reach the sphere of the neuter. But nothing is less certain, as I have already tried to show and as Esposito himself shows regarding Levinas.[34]

The crucial importance of Agamben's reading of Heidegger's 1929–30 seminar is the linkage he establishes between Heidegger's notion of life and his own notion of the political, and the step forward taken on the association of Deleuzian beatitude (which is toward the end of *The Open* conflated with the Benjaminian image of a nature reconciled, a "saved night"[35]) with Heidegger's boredom as fundamental attuning. Agamben succeeds, through Heidegger, in projecting Esposito's critique of the ideology of the person and his proposal for a politics of the impersonal onto productive practical terrain up to a certain point.

Agamben directly links the understanding of life in *Being and Time*, which it is the function of the *World, Finitude, Solitude* seminar to explicitate, to contemporary biological and zoological work (Hans Driesch, Karl von Baer, Johannes Mueller, and Jakob von Uexkuell are mentioned[36]). Heidegger's fundamental insight, already from *Being and Time*, is that "the ontology of life is achieved only by way of a privative interpretation";[37] "privative" is meant in the sense of a subtraction and a reduction to mere present-at-handness. The concept of life inherited from the tradition is therefore thoroughly anthropocentric. The 1929–30 seminar will reconduct the investigation of both the abyss and the peculiar proximity between *Dasein* and animals towards a sphere where "*animalitas* become[s] utterly unfamiliar and appear[s] as 'that which is most difficult to think,'" and "*humanitas* also appears as something ungraspable and absent, suspended as it is between a 'not-being-able-to-remain' and a 'not-being-able-to-leave-its-place.' "[38]

If it is true that "the animal behaves within an environment but never within a world,"[39] this is so because, according to Heidegger, animals and humans have a different relationship to the open. Animals are "open in a nondisconcealment," that is, they are captivated by what calls them. If the openness of humans to the world is primarily an openness to the conflict of truth, which is the conflict between concealment and disconcealment, then it follows that "the openness of the human world . . . can be achieved only by means of an operation enacted upon the not-open of the animal world."[40] Boredom is the "place of this operation," which immediately places boredom, as abandonment in emptiness, as the essential phenomenal hinge tracing the human/animal separation. In other words, boredom is the key to the overcoming and dismantling of the bare-life *dispositif* to the same extent that it holds the key to an understanding of the difference between environment and world.

"In becoming bored, Dasein is delivered over to something that refuses itself, exactly as the animal, in its captivation, is exposed in something unrevealed";[41] hence, "both [animals and humans] are, in their most proper gesture, open to a closedness; they are totally delivered over to something that obstinately refuses itself."[42] The *lethe* (concealment) is primary in the unconcealment of *aletheia* as truth. The passage from environment to world is the disruption of animal captivation, but this immediately means: "the open and the free-of-being do not name something radically other with respect to the neither-open-nor-closed of the animal environment: they are the appearing of an undisconcealed as such, the suspension and capture of the lark-not-seeing-the-open."[43] This sets the caesura between animals and humans into a structure of mutual belonging: the conflict of truth, the originary and ceaseless struggle between concealment and unconcealment, now emerges as the originary and ceaseless struggle between environment and world, between animals and humans. And "Dasein is simply an animal that has learned to become bored; it has awakened *from* its own captivation *to* its own captivation."[44] This awakening (awakening to a closedness) is the rise of the human, which sets the human as not just the result but the infinite production of the awakening, which is also at the same time, as in boredom, or as in the struggle for truth, the infinite suspension of the awakening.

This is indeed the beginning of politics, even of politics for a technopolitical age: politics as awakening, politics as the pursuit of a path of disclosedness against the fundamental occlusion of the possible. "The ontological paradigm of truth as the conflict between concealedness and unconcealedness is, in Heidegger, immediately and originarily a political paradigm. It is because man essentially occurs in the openness to a closedness that something like a *polis* and a politics are possible."[45] But the very possibility of a politics does not mean its goodness. Heidegger himself pursued what he thought of as a good politics in terms of a reconstitution of community alongside the production of a we. For a number of years, he thought that a personalization of the political possibilities of truth was not just possible but in effect historically destined in terms of the authentic possibilities of the German people. At the same time, and even while remaining blind or captivated by his own investment, he warned against another move of the sinister, which he associated with the destiny of the entire West under metaphysics: the return of disawakening under the sway of technology.

If in Heidegger's position we can recognize a reconceptualization of the political that makes it thoroughly dependent on the struggle of the human against animality; or in other words, if from a Heideggerian perspective we can see that politics is always already biopolitics, then two paths seem to open up for us. One of them is the recaptivation of the human by its own pretenses: the human that has left the animal behind, the human that has thoroughly personalized and reified its own awakening into an ideology of endless production of world, of the ceaseless management of life, increasingly as technical world, and as technically administered life, is also the human that, in Agamben's words, "seeks to open and secure the not-open in every domain, and thus closes itself off to its own openness, forgets its *humanitas*, and makes being its specific disinhibitor. The total humanization of the animal coincides with a total animalization of man."[46] This is the biopolitical project, relentlessly pursued today by liberal humanity and its absolute reliance on technology: it is also, inevitably, an impersonal politics of world production as world domination. The night of the world threatens in that path, and it is not the saved night. It is a night because it is the return of the undisclosedness of world under the sign of a radical (technopolitical) disawakening. If a conciliation of humans and animals is to be found there, it will be at the price of a thorough extension of the apparatus of depersonalization in the direction of subjection. There will be no way of inverting the sign of the dominating force into an active force of redemption. No attempt at inverting the sign of biopolitics into an affirmation of the *lethe* at the heart of the living can go beyond humanitarian ideology, which is, as Esposito has shown in his deconstruction of the person, not the obstacle but rather one of the reasons for the present sway of biopolitics and the consolidation of the bare-life apparatus, beyond ostensibly political life, into the technocapitalist constitution of postgenomic life and the thorough control of the organic (and this is the other side of nano-technology's promises).[47] Something else is needed.

And this might be the second path that, according to Agamben, opens after Heidegger. Agamben mentions it: "man . . . appropriates his own concealedness, his own animality, which neither remains hidden nor is made an object of mastery, but is thought as such, as pure abandonment."[48] To think the pure abandonment of captivation, and to elaborate a practical politics on that basis, is perhaps not to be understood through the notion of appropriation. For man to appropriate his own concealedness means: for man to reconstitute himself as a subject, even as a personal subject (since there is no other kind). Resubjectivation is the limit of an affirmative biopolitics, Agamben suggests, which is perhaps the reason why Esposito ends his book with a call for the constitution of a *persona vivente*: "not separated from, or implanted in, life, but coinciding with it as unsplittable conjunction [*sinolo*] of form and force, outside and inside, *bíos* and *zoé*" (*TPN*, 183–84). But, has beatitude not become too much here? Has *homo tantum* not dangerously crossed the border of separation between the infinite production of awakening and the infinite suspension of awakening into, once again, the thematic disawakening of a univocity in being? Can the *persona vivente* do anything but at most offer itself as an ideological banner to ward off the most egregious abuses against disposable human life? This is finally my question for Esposito, and also for Agamben.

Why define being as force? Force is force. A politics of the impersonal that can move beyond interhuman relations into the animal and the organic, and vigilant enough to harness the force of technopolitics and to refuse to become one with it, is not necessarily, not in principle, a politics of conjunction or conciliation nor does it need to be. If there is one thing that politics cannot do, it is to be less than political. Does attention to the impersonal warrant a politics of life? In the affirmation of "life" as the horizon of a philosophy of the future, does life not become yet another transcendental illusion? The historical captivation of the human is a non-cell-based conspiracy, as Thomas Bernhard puts it in the epigraph to this essay. Why should the animals' captivation be different? At this point of maximally troubled proximity, a technopolitics commensurate to technoscience should ask itself that question.

An Infrapolitical Solution

The dead center of Esposito's 2016 *Da fuori* is the presentation of the scission between deconstruction and biopolitics as decisive for contemporary thought. This is explicitly an interested statement, to the extent Esposito will end up claiming his own position at the mediating point of both perspectives (*DF*, 180).[49] At the same time, he also claims that an affirmative biopolitics can no longer share the post-Heideggerian paradigm of deconstruction (*DF*, 171). Esposito will in effect place the central conflict of his book as a conflict between deconstruction and biopolitics. Deconstruction will have been on the side of "French Theory," and biopolitics on the side of "Italian Thought." The third contender in this gigantomachy is "German Philosophy," a little worse for wear but acknowledged, particularly in its Adornian avatar, as the occupier of the structural site of negativity. German negativity would have been historically vanquished by the impolitical "neutralization" of deconstruction (and of thinkers such as Jean-Francois Lyotard, and of certain not-too-fruitful aspects in the thought of Deleuze or Foucault). An Italian thought of the outside would have come to take over from French neutralization through a radical affirmation, which is the symbolic position where Esposito sees himself, aided by explicit developments of the thought of life in Deleuze and Foucault through an intensification of their Nietzschean genealogy.

Esposito makes his own position in Italian Thought complex by finally rejecting the legacy of political theology whose investigation has characterized so much of the best contemporary Italian reflection (Mario Tronti, Massimo Cacciari, and Agamben among others). Esposito would be looking, through his affirmative biopolitics, for a destruction of political theology. Esposito asks, and leaves his answer implicit, whether the destruction of political theology would immediately mean the end of politics, that is, the impolitical neutralization of politics. Esposito's wager is not for an end of politics, rather it is for a militant politicality, which is in *Da fuori* exemplified in his proposal for a concrete politics concerning Europe in its present conundrum. He calls it "grand politics:" "Never as now, in a present in continuous flight towards an uncertain future, there has been such a vital necessity of what was once defined as 'grand politics'" (*DF*,

226). The attempts at the redefinition of Europe as a "civil power" in Italian jurisprudence are reconducted by Esposito toward the notion of a "popular power," with a basis in the thought of Machiavelli and Giambattista Vico. If, for Machiavelli, the great seek to oppress and the people seeks not to be oppressed, also for Vico the people, far from forming an undifferentiated and homogeneous set of individuals, is constituted by and through the conflict itself, "a social class opposed to another class" that antagonizes it (*DF*, 237). As a popular power Europe would think of its grand politics as invested in the formation of a people, understood as the formation of a class that refuses to be oppressed and constitutes itself in antagonism to the oppressor class: "Europe's process of political unification . . . will be the result of a real political dialectic. It will no longer respond to the politico-theological machine that imprisons our lives, it will rather work towards its dismantling" (*DF*, 238). Yes, this is a grand-political project for Europe, but at the same time its affirmative biopolitics component gets diffuse: it is more claimed than shown.

There is no nationalist interest in Esposito's claim for Italian Thought. Any nationalist interest would be immediately undermined by the thematization of the outside that Esposito posits as the privileged instance of thought in the last century. If biopolitics is today at the center, it is because it can install itself in the outside affirmatively, in view of political efficacy, which is what the "Germans" and the "French" were never able to reach. Frankfurtian negativity is as politically deluded as French neutralization, which moves toward an impolitical maximization. Affirmative biopolitics would be in a position to close successfully, that is, productively, the great crisis of thought intimated by Husserl and Heidegger, Valéry and Ortega, Wittgenstein and Spengler. But this seems too large a claim.

In its world avatar (that is, in its importation to Anglo-Saxon universities, and beyond) French Theory, which was already a reaction to German Philosophy, releases itself into a literaturization of thought that deprives it of the possibility of antagonism to every form of the real. It is a problem that only increases with the North American reception. Literaturization means impolitization, in a complex way. In reference to Derrida, Esposito says that, on the one hand, "the politics implicit in his texts is born, not so much from a theory of politics, but rather from a kind of resistance in its comparisons and confrontations—from a kind of allergy to any normative philosophy. In that sense it is not possible to speak of the depoliticization of a discourse which, like that of Derrida, is endowed with an immanent politicity, implied in deconstructive practice"

(*DF*, 135). On the other hand, he claims, "it is not possible either to sustain that Derrida's philosophical research carries an effective political gesture" (*DF*, 131). That is, for Esposito, on the one hand, the politicity of deconstruction is immanent to the deconstructive gesture that rejects any normative politicization. On the other hand, it cannot reach effective politicality. It remains, after all is said, literature.

But one must wonder whether Esposito is passing on to deconstruction a notion of politics that deconstruction will have rejected beforehand. The notion of the impolitical that Esposito uses, referred of course to Esposito's 1988 book *Categories of the Impolitical*, becomes insufficient: "[Derrida's] more recent works also . . . remain as impolitical as the first. And that, notwithstanding the evident efforts of the author . . . to assume as his object political categories, such as that of democracy" (*DF*, 131). Esposito begins to force his argument in this contradiction in his characterization, which would seem to reduce all politicality to a principle of action, and which can only be solved in the attribution to Derrida, and to the totality of French Theory, of a "logic of neutralization" of every practical decision (*DF*, 131). This means, in Esposito's explanation, to remain fixed in a logic that excludes conflict and that remains "this side of the line of politics" (*DF*, 131). But with this Esposito engages in a reduction that exceeds his presuppositions. He says that in Derrida or from Derrida "one can no longer distinguish friend from enemy, or set into practice any link that is not always already delinked" (*DF*, 131). To the "politics of separation" Derrida invokes in *Politiques de l'amitié* (1994) Esposito responds with a "separation from politics" (*DF*, 131), which seems to me an illegitimate gesture. Obviously, we must read it retrospectively, from the end of the book and the grand-political approach taken there, as a denunciation of any possibility for deconstruction to establish itself politically within the Machiavellian-Vician universe of conflict in favor of a "popular power" that could reject oppression. Deconstruction, says Esposito, cannot take sides, and remains suspended in a neutralizing netherworld.

But there is no separation from politics in the politics of separation. It could also be said that the politics of separation summons forth a hyperpolitization, repoliticizing as such, which is far removed from any politico-theological presuppositions, including the current substitution of the concept of effective hegemony for that of the general will. A politics of separation is necessarily posthegemonic. At least it is so understood by infrapolitical reflection, from the side of deconstruction and from the post-Heideggerian side, but also from the subalternist and postcolonial side. A politics of separation is a radical practice of parrhesia that refuses every attempt at submission of the singular existent to any instance of order and to any communitarian instance in itself turned autoimmunitary. As an example, we can see these days, in places where a "new" politics has made itself potentially present on the side of the left, the danger (luckily not come to fruition yet, hence the political importance of warding it off) of the sustained appeal to the faithful in favor of the unity of and trust in the forces of change. In such cases it is argued that it is desirable, even imperative, to commune in the constitution of a closed community of militancy whose effective model might perhaps be, in spite of everything, and in spite of the 15-M movement, for instance, still the old-fashioned Leninist-Stalinist communist party. But the constitution of a closed community of militancy always places the community in an autoimmunitary predicament where the community labors for its own destruction: in the communitarian or party closure in favor of the unity of positions, in favor of hegemonic constitution, politics is suspended and only faith is affirmed. A politics of separation is openly against that, and not only in conditions of constituted power, where silence and conformity are requested from already formalized institutional sites. It is also against it, and should be, under constituent conditions, where silence and conformity and the unbreakable allegiance to the leader or leaders carry with them the promise of the militant suspension of politics and undermine from within every promise of effective emancipation from power, even if power is now to be defined as a new counterhegemonic power—communal power or the power of the multitude. There is no literature there, but to be against it is not literary either. It is rather a matter of a non-impolitical hyperpoliticization, subtracted from politico-theological conditions of existence, and certainly well away from any affirmative biopolitics, if the latter were to imply the principial celebration of communitarian life as such, of communion as political procedure. But we must see in detail how Esposito argues his case, in order not to be unfair to him. For now let me note that my argument repeats what Giorgio Agamben says about the polarity constituent-constituted power, which Esposito otherwise sanctions.

Discussing Agamben's "messianic perspective," and comparing it with Tronti's "eschatological perspective" and Cacciari's "katechontic perspective," and wondering whether the end of political theology will necessarily imply an end of politics, Esposito alludes to the passages in *L'uso dei corpi* (*The Use of Bodies*) where Agamben gives up on the constituent-constituted power bipolarity, denouncing it not just as intrinsically politico-theological but as doomed to inefficacy (everything constituent becomes constituted, everything open closes itself, everything liberating will end up oppressing, and this is a function of the call to the constitution of a power in the first place). Agamben says: "for destituent potential it is necessary to think entirely different strategies, whose definition is the task of the coming politics. A power that has only been knocked down with a constituent violence will resurge in another form, in the unceasing, unwinnable, desolate dialectic between constituent power and constituted power, between the violence that puts the juridical order in place and violence that preserves it."[50] But Agamben is, for Esposito, still looking for "a different figure of politics" (*DF*, 193). Agamben calls it "destituent potential." Esposito's critique is that destituent potential is already internal to the politico-theological *dispositif*: "the 'end' of political theology is [in Agamben] evoked in a politico-theological language, in this case of a messianic character" (*DF*, 193). It is necessary to take another step out of political theology which cannot be taken from the subordinate (that is, still internal to political theology) category of secularization, or from any of its variants (disenchantment, profanation): "the 'resolution,' in the literal sense of the term, of political theology does not go through the category of secularization. Just as it does not go through the categories of 'disenchantment' or 'profanation,' from the moment that the latter place themselves on the reverse side of what they claim to contest. They are, that is, necessarily trapped in the dialectic that joins them, negatively, to enchantment and the sacred" (*DF*, 194). Against the messianic-secularizing position of Agamben, Esposito calls for a further critical step, in the understanding that transgression, as Georges Bataille made clear in his own work, does not destroy the law but only maintains it and confirms it. It seems difficult to imagine Agamben himself would have been blind to this argument. If it has not escaped him, then it is easy to conclude that Agamben has no particular intention of taking political theology to its end. He needs it. I wonder whether a similar case could be made for Tronti or Cacciari. Esposito affirms his freedom from such a problem, but he does it against the literaturizing neutralization of impolitical deconstruction.

The Pauline "*hos me*" ("as if not"), which is in Agamben the mark of his destituent position, since it unworks the law in faith and opens a new figure of the political in its secularized, profane version, is to my mind effectively deluded as a figure of politics and therefore also as the figure of a new politics. Agamben, perhaps unnecessarily, errs in his insistence on the formation of a new politics. But is Esposito not doing the same, in a different manner? In the chapter devoted to "Italian Thought" Esposito proposes as his central thesis the notion that Italian thought is "the attempt to confer political form" on what, in Adorno's negative dialectics, in the "dynamic between power and resistance" Foucault theorized, or in the "dichotomy between the 'molar' and the 'molecular' " in Deleuze, not to speak of the allegedly excessive emphasis on death in the Derridian paradigm, for which life is essentially survival, "rests on an inevitably impolitical plane" (*DF*, 170). The "matter" of Italian philosophical discourse will no longer be the social, or writing, or the "neutral circuit of an equivalence between forces" (*DF*, 171), but rather "the political apprehended in its inevitable conflictual dimension" (*DF*, 170). In contemporary Italian reflection, Esposito tells us, there are three conceptual bipolarities that move thought toward its radical politicity and politicization. The first two refer to the early work by Mario Tronti on workers and capital, and to Negri's meditation on constituent power. The third refers to Esposito's own work: to community and immunity. For Esposito, the sense of community registered in his earlier work does not rely on any notion of identity, but rather on a "constitutive alterity" that refers to an "exteriorization" and to a "reciprocal contamination" (*DF*, 179). Esposito's emphasis on *munus*, however, rather than on the *cum* emphasized by the sort of deconstructive expropriation of community apparent in Jean-Luc Nancy, for instance (*DF*, 180), would have enabled Esposito to radicalize the thought of community toward its political valence. *Munus* is common to community and to immunity: "If *communitas* links its members in a reciprocal pursuit, *immunitas* exonerates them from such a duty. In the same way that community remits to something general, immunity remits, on the contrary, to the privileged particularity of a condition subtracted from communitarian obligation" (*DF*, 181). And this latter thing is the problem. Today, for Esposito, we would have amply overtaken the modern immunitarian need, we would have exceeded the limit before which immunity can still be proposed or fought for as effectively political, and we would have reached a stage where immunity, and with it every kind of immunitarian emphasis, would be "no more than a cage where not just freedom consumes itself but also the very sense of existence—that opening to the outside that has been named community" (*DF*, 181).

This is, I think, the place where all the previously woven threads of my critique of Esposito's affirmative biopolitics come together. For Esposito—retrospectively, one could perhaps already see him coming in this respect since the last few pages of *Third Person*—community comes to be understood as the central, essential site of politics for our epoch. Community is even the promise of an epochal overcoming of immunitary nihilism. Today, for Esposito, "the common has become the real and symbolic form of resistance to the excess of immunization that captures without an end" (*DF*, 181). Italian Thought must therefore be defined, in its apex, as communitarian thought, from an understanding of community as the active force in respect of which immunity is reactive. From here we can understand the preceding history of reflection, that is, modern reflection, as the elaboration and circulation of "a series of narratives oriented to providing increasingly effective immunitarian responses against the risks, real and imagined, of human relations. The modern political is not characterized as such by the norm of the exception, by inclusion or exclusion, but by the immunization of a life deprived from transcendent protection and delivered to itself" (*DF*, 183). Affirmative biopolitics, committed to a reactivation of the common, must however face the absence of jurisprudence turned toward the institution of the commons, the lexical absence regarding the commons, and the absence of conceptual categories from which to think the commons—since for many centuries the immunitary paradigm and "its politico-theological language" have organized our world. The practice of the outside, if it could in the past have been the very opposite, is today communitarian practice, understood as the search for the deactivation of immunitary practices such as those dependent on the privilege of the upper classes, which can only solve themselves in the oppression of those who would rather not be oppressed. This is Esposito's proposal for a grand affirmative biopolitics, which indeed reformulates and orients the Machiavellian position, and indeed sets forth, if not a program of action, at least the justification for it.

Returning, however, to the scission between deconstruction and biopolitics which organizes the critical point of departure of *Da fuori*, it remains to examine up to what point deconstruction can only be read in an immunitary and countercommunitarian key. I suppose Esposito places deconstruction as the favored enemy because he perceives it as the other thought tendency that has also left behind the language of political theology. If I can be somewhat reductive, if deconstruction and affirmative biopolitics are, in Esposito's version, the only two visible instances of thought where the politico-theological architecture seems to have been demolished, then it is all the more important to eliminate the immunitary and neutralizing residue of deconstruction.

This is the reason why communitarian thought must take its point of departure in a fundamental rejection of the basic infrapolitical position, which *Da fuori* presents through some words by Nancy: "Thought does not dictate and does not guarantee what must be decided or that it is decided. That is its archi-ethics and its specific responsibility" (*DF*, 133). An elaboration of what exists behind those words by Nancy would belie the notion that deconstruction must be thought of as contained in the immunitary paradigm. Deconstruction is, rather, a subtraction from what Esposito calls the "logical precedence" of community with respect of immunity. Not that it would prefer to invert the precedence: the subtraction is a subtraction from the very choice between the two, from the very decision and its dictates and guarantees. In the affirmative biopolitics paradigm, community must rule today over any attempt at immunitary activation. But the obvious question is, until when? If modernity can be defined as the history of immunitary excesses against community, at what point of a new effective politics will we feel obliged to state that communitarian excess will overflow, has already overflown, its own conditions and threatens the necessary immunity of the singular existent, without which there would be not just no freedom but also no sense of existence?

Da fuori does not quite ruin deconstruction. On the contrary, deconstruction remains as the effective limit, hyperpolitical and anything but neutralizing, and precisely for those reasons also infrapolitical, in the face of a political appeal that, no matter how grand it means to be, can still not give an adequate account of its own conditions of enunciation. In the book that precedes *Da fuori* in Esposito's production, *Persons and Things*, Esposito ends by sending us off into an unresolved nightmare. If, for Esposito, the fundamental political phenomenon of our time is the fact that "immense masses multiply themselves in the squares of half the world" and "their words incarnate in bodies that move in unison, with the same rhythm, within a single affective wave," if "still deprived of adequate organizational forms, bodies of women and men push the borders of our political systems, asking for transformations irreducible to the dichotomies that have for so long produced the modern political order"[51] (*PC*, 110), that so-called "radical novelty" of the "living body of the multitudes" (*PC*, 111) (remember the notion of "*persona vivente*" that concluded *Third Person*) is not only asking for a new communitarian lexicon, but is also in dire need of a response capable of interrupting the exclusionary link between politics and community—something that affirmative biopolitics seems to have some trouble producing. But that is, after all, the function of the politics of separation Derrida once called for.

Notes

1. On Heidegger's notion of "mere life" (*Nur-Lebenden*), which Derrida must have associated with Agamben's work, this is what Derrida says: "I think I understand what that means, this 'nothing more' (*nur*), I can understand it on the surface, in terms of what it would like to mean, but at the same time I understand nothing. I'll always be wondering whether this fiction, this simulacrum, this myth, this legend, this phantasm, which is offered as a pure concept (life in its pure state—Benjamin also has confidence in what can probably be no more than a pseudo-concept), is not precisely pure philosophy become a symptom of the history that concerns us here" (Jacques Derrida, *The Animal that Therefore I Am*, ed. Marie-Louise Mallet, trans. David Wills (New York: Fordham University Press, 2008), 22). A reading of *The Animal That Therefore I Am*, Derrida's posthumous and unfinished book on subjects that are very close to Esposito's and Agamben's, would be advisable, but I must renounce that task for reasons of space.

2. Giorgio Agamben, "Absolute Immanence," in *Potentialities. Collected Essays in Philosophy*, ed. and trans. Daniel Heller-Roazen (Stanford: Stanford University Press, 1999), 238.

3. As Deleuze glosses it, "a scoundrel, a bad apple, held in contempt by everyone, is found on the point of death, and suddenly those charged with his care display an urgency, respect, and even love for the dying man's least sign of life. Everyone makes it his business to save him. As a result, the wicked man himself, in the depths of his coma, feels something soft and sweet penetrate his soul. But as he progresses back towards life, his benefactors turn cold, and he himself rediscovers his old vulgarity and meanness. Between his life and his death, there is a moment where *a life* is merely playing with death" (Gilles Deleuze, "Immanence: A Life . . . ," in *Two Regimes of Madness. Texts and Interviews 1975–1995*, ed. David Lapoujade, trans. Ames Hodges and Mike Taormina (New York: Semiotext(e), 2006), 386).

4. Deleuze, "Immanence: A Life . . . ," 387.

5. Ibid., 385.

6. Agamben, "Absolute Immanence," 225.

7. Deleuze and Guattari quoted in Agamben, "Absolute Immanence," 227.

8. Agamben, "Absolute Immanence," 233. Regarding necessary illusions, Agamben says: "This illusion [of transcendence] is . . . something like a necessary illusion in Kant's sense, which immanence itself produces on its own and to which every philosopher falls prey even as he tries to adhere as closely as possible to the plane of immanence" (Ibid., 227).

9. Ibid., 239.

10. Ibid., 233.

11. Ibid., 236.

12. Agamben refers to the Sartrian *il y a* as developed in "*La transcendence de l'ego*," which Deleuze studies in *Logic of Sense* (Agamben, "Absolute Immanence," 224); and Esposito makes reference to the importance of the alternative Levinasian *il y a* for Blanchot (*TPN*, 158).

13. Deleuze, "Immanence: A Life . . . ," 388–89.

14. There is of course a lot to argue here if Agamben implies that Heidegger, Levinas, or Derrida fall into transcendental subjectivity as victims of the Kantian "necessary illusion."

15. The notions of "active and reactive forces" are of course major interpretive categories in Gilles Deleuze's *Nietzsche and Philosophy* (New York: Columbia University Press, 1986), 39–72. I have always felt that the attribution of those categories as ultimate ontological categories to Nietzsche is excessive, although they show up in many guises in Deleuze's thought—they belong more to Deleuze than they do to Nietzsche.

16. All the English translations from *Terza persona* in this paper are mine.

17. It is however arguable that the ontotheological primacy of the concept of the person did suffer a great challenge precisely through the history that Esposito retraces in the first chapter of *Third Person*, namely, the period between the development of modern biology in the early nineteenth century and the end of World War II.

18. Esposito makes it clear that for the long tradition of civil law the human body is juridically not to be confused with the thing. The body is not what one has, but rather what one is (see *TPN*, 114–18).

19. The essay in question is "*La nature des pronoms*" (1956), published in *Problemes de linguistique generale* (1966).

20. Esposito refers to Agamben's *The Open* punctually, to state that "through different paths, it reaches the same conclusions" (*TPN*, 140, n.19). My purpose will be to expand on those conclusions.

21. Hiu Lui Ng was a New Yorker of Chinese nationality, married to an American citizen and with two American-born sons, who was arrested in the process of his final interview for a green card on the basis of having overstayed his visa years earlier. He was detained and shuttled through jails and detention centers for many months and denied urgently necessary medical care until he died in custody after tremendous physical suffering. The withholding of medical care is another aspect of the technopolitical administration of life, especially for people deemed to be non-persons or quasi-persons or less-than-persons (see Nina Bernstein, "Ill and in Pain, Detainee Dies in U.S. Hands," *New York Times* August 13, 2008, A1; A16).

22. For "immunitarian drift" and in general for the notion of immunity see Esposito's *Immunitas*.

23. Paul Preston, *Comrades! Portraits from the Spanish Civil War* (London: HarperCollins, 1999), 129.

24. About the non-subject see Alberto Moreiras, *Línea de sombra. El no-sujeto de lo político* (Santiago: Palinodia, 2007).

25. Giorgio Agamben, *The Open. Man and Animal*, trans. Kevin Attell (Stanford, CA: Stanford University Press, 2004), 3.

26. Ibid., 6.

27. Ibid., 12.

28. Agamben links Deleuzian beatitude with the Gnostics, and then with Benjamin, as part of the underlying eschatological frame of his book (see Agamben, *The Open*, 81–84). The notion of "the whatever" refers to Agamben's *The Coming Community.*

29. "When the difference vanishes and the two terms collapse upon each other—as seems to be happening today—the difference between being and the nothing, licit and illicit, divine and demonic also fades away, and in its place something appears for which we seem to lack even a name" (Agamben, *The Open*, 22).

30. Ibid., 29. It is interesting that the restitution of the notion of human nature in a substantive sense is for conservative thought in general the best defense against technoscientific threat (see Francis Fukuyama, *Our Posthuman Future. Consequences of the Biotechnology Revolution* (New York: Picador, 2002), 29–47).

31. Agamben, *The Open*, 37, 38.

32. Ibid., 16.

33. Agamben, "Absolute Immanence," 239, 225.

34. And as I would have liked to have shown also for Derrida, whose book on "the autobiographical animal," mentioned above, is a radical critique of the separation human/animal in the philosophical tradition and down to the very same Heideggerian text Agamben focuses his analysis upon. See below.

35. Agamben, *The Open*, 81.

36. Ibid., 51.

37. Heidegger quoted in Agamben, *The Open*, 50.

38. Agamben, *The Open*, 50–51.

39. Heidegger quoted in Agamben, *The Open*, 51.

40. Agamben, *The Open*, 62.

41. Ibid., 65.

42. Ibid.

43. Ibid., 68.

44. Ibid., 70.

45. Ibid., 73.

46. Ibid., 77.

47. Among the many possible books on these subjects see Jürgen Habermas, *The Future of Human Nature* (Cambridge, UK: Polity Press, 2003); Fukuyama, *Our Posthuman Future*; Kaushik Sunder Rajan, *Biocapital. The Constitution of Postgenomic Life* (Durham, NC: Duke UP, 2006).

48. Agamben, *The Open*, 80.

49. All the English translations from *Da Fuori* in this paper are mine.

50. Giorgio Agamben, *The Use of Bodies*, trans. Adam Kotsko (Stanford: Stanford University Press, 2016), 266.

51. All the English translations from *Le persone e le cose* in this paper are mine.

5

Fold of Life

Roberto Esposito on "the Living Person" and Animistic Personhood

Inna Viriasova

On the last page of his book *Third Person: Politics of Life and Philosophy of the Impersonal*, Roberto Esposito evokes the notion of "the living person" (*la persona vivente*), concluding his thorough genealogical analysis and deconstruction of the modern *dispositif* of the person. He writes: "the *living person*—not separate from or implanted into life, but coextensive with it as an inseparable *synolon* [a composite whole] of form and force, external and internal, *bios* and *zoé*" (*TP*, 151). In this concluding notion, Esposito seeks to reconcile life (force, interiority, *zoé*) with personhood (form, exteriority, *bios*), which have been torn from each other throughout modernity. However, this conclusion is not the final word but an opening, insofar as "the living person" remains more of a hint than a developed concept, a "figure that has yet to be fathomed," as Esposito puts it. Speaking to this opening, this chapter will first sketch Esposito's critique of the *dispositif* of the person and then suggest how "the living person"—a paradoxical notion, considering Esposito's endeavor to rid our discourse and practice of the primacy of the personal—can push current political philosophy beyond its long-standing determination by the horizon of naturalism, responsible for the exclusionary and reifying logic of personhood.

According to Esposito, a challenge to the *dispositif* of the person surfaces under various names in several twentieth-century philosophies of the impersonal; in particular, he turns to Weil, Benveniste, Kojève, Jankélévitch, Levinas, Blanchot, Foucault, and Deleuze to provide "a preliminary theoretical grid" for a philosophy of the "third person." While Esposito's analysis of all the figures of the impersonal is rich and full of potential applications, I want to focus on the particular challenge that "life" posits to the *dispositif* of the person.[1] In this regard, we can find the clearest elaboration of "the living person" in Esposito's engagement with the impersonal figure of "life" in the philosophies of Foucault and Deleuze. I suggest that the opening, which the notion of the living person performs here, is twofold. First, from a Foucaultian perspective, it is indicative of the ever-present emergence of new forms of living resistance that traverse the modern *dispositif* of the person and thus lay the ground for "an *affirmative* biopolitics." Second, Esposito's reading of Deleuze on "*a* life" and "becoming-animal" implies a slightly different opening that has a potential to traverse not only the repressive biopolitical regime, but any form of politics whatever grounded in the naturalist ontology of Western modernity. In this respect, "the living person," coextensive with the life that animates it, is a figure of the immanence and the indivisibility of life or, better still, an alternative *dispositif* or "regime of meaning" that produces real effects at the level of life itself, prior to its problematic relationship with politics and law. This new *dispositif* of the *living* person takes political philosophy toward a different horizon, one that, under a general name of "animism" or "new animism," has recently become an important tool of "decolonization of thought"[2] within such varied academic domains as anthropology, geography, political theory, and philosophy. Esposito's philosophy of the living person, I suggest, can be situated within this general movement toward "animistic ontology" in contemporary political thought. As a result, we might further develop the notion of the living person by turning to the multiple indigenous conceptions of personhood that rely on *contamination* rather than immunitary sanitation in the process of composition of human and non-human persons.

The Apparatus of Separation and Exclusion: Human and Animal, Person and Thing

The main conclusion of Esposito's genealogy of the person can be summarized as follows: the political ordering of living beings and things in modernity relies on the mechanism of distinction and separation between what is considered the *natural* and thus *impersonal* layer of biological existence and the *artificial* domain of *personhood.* This distinction equally applies to the individual and the species, producing division within the unity of a living being and within its community with other beings. Even though the hierarchy between these two domains (nature and artifice) has shifted throughout history, the distinction between them has remained intact. Esposito calls this mechanism of division "the *dispositif* of the person," borrowing the term from Foucault's conceptual toolbox. The notion of *dispositif,* alternatively translated as *apparatus,* conveys the "performative role" of the idea of the person, meaning that it is necessarily productive of real effects, subjects, and subjectivities, and relies on multiple ensembles of material practices, discourses, and institutions. In this way, the person is not just a philosophical idea but an ongoing practice of ordering and control (*TP,* 9).[3] Most importantly, Esposito argues, it is a practice of not only division but also active exclusion, the constitution of the limit between the inside and the outside, "of inclusion through the exclusion of what is not included."[4] The constitution of the person, as the proper space of belonging and interiority, is based on a separation from exteriority; and this logic is preserved when applied to a group—for there to be *us,* there has to be *them.* Interestingly, this exteriority or otherness does not always lie outside an individual or a social body but, as Esposito shows, is often located within the confines of the same self: the living being is split from within and humanity as species is divided against itself. Esposito's philosophy of the impersonal is situated against this background of separation inherent in the modern *dispositif* of the person, and the notion of the *living* person, as I will suggest below, contrary to the person—an artificial mask implanted upon the natural face of a living being—is a different kind of mask that restores the unity of life with itself and empowers its further transformation.

Esposito argues that an important shift in the philosophical paradigm of modernity occurs in the early 1800s, when developments in biology produce a crisis in the existing understanding of the concept of the person. Grounded in Roman legal tradition and Christian theology, for centuries the person has been considered the locus of rational action, will, and responsibility, the aspect of an individual self that was truly human and in control of its natural or animal aspect. This personal self has been the object of legal imputation as well as the subject of a promise of resurrection, and as such was distinct from and in excess of the body to which it was attached. This primacy of the personal over the impersonal or biological self was shaken by Xavier Bichat's theory of life's two modes of being, which very soon found its way into philosophy as well as sociology, anthropology, zoology, and linguistics. This event, importantly, also marks the establishment of biopolitics as the primary form of modern governmentality.

The novelty of Bichat's definition of life lies in its essential juxtaposition with death; as he famously put it, life is "the sum of the functions, by which death is resisted."[5] Through his medical practice Bichat discovered that there are two "kinds" of death and, correspondingly, there are two modes of life: "organic life," which includes such vegetative functions as digestion, respiration, and circulation of the blood, and "animal life," which governs the motor, sensory, and intellectual activities. As Esposito sums up, "[w]hile organic life is closed and inward looking, animal life is in contact with the environment, changing it and being changed by it" (*TP*, 22). Importantly, Bichat asserts the functional and quantitative prevalence of organic life over animal life, insofar as the former appears before and also persists after the latter (e.g., the nutritive life of the fetus before birth and the "growth" of nails and hair after death) (*TP*, 23). The effects of such a doubling of life are detrimental to the modern understanding of political subjectivity and thus to the concept of the person: "[i]t is as if a non-human—something different from and earlier than animal nature itself—had taken up residence in the human being; or as if it had always been there, with dissolutive effects on the personal modality of this being." This primacy of vegetative or non-human life within the human implies that "a political subject [the person] does not even exist as a source of voluntary action, because the will, although linked to animal life, is deeply innervated in a bodily system that is sustained and to a large extent governed by its vegetative part" (*TP*, 24). Consequently, if social-political actions are determined by biology, politics needs to focus on the government of the biological given of its living subjects, transforming inevitably into *biopolitics*. The pinnacle of such a total collapse of the rational or personal part of the human into its biological existence was Nazism with its thanatopolitical horizon. However, the problematic aspects of biopolitics are not limited to its extreme historical instances like Nazism, but resurface every time a distinction is made between the human and the inhuman, replicating Bichat's doubling of life inside the human as well as within humanity as a whole.

The task of delimiting and separating humanity from the non-human or, more generally, "culture" (properly human domain) from "nature" (neutral domain shared by all beings) is paradigmatic of modern "naturalist ontology." In this regard, in his engagement with Bichat's thought, Esposito, I believe, alludes to a more general horizon of "naturalism" that has defined Western modernity. According to contemporary French anthropologist Philippe Descola, in naturalist ontology the continuities and discontinuities between various beings are established based on the selective attribution to them of "interiority" (what is generally called "the mind, the soul, or consciousness") and universal attribution of "physicality" (such as "external form, substance, the physiological, perceptive and sensorimotor processes").[6] This naturalist distinction between inside and outside is analogous to Bichat's differentiation between organic and animal life: while organic life or physicality is shared by all beings (human and non-human, organic and even, perhaps, non-organic, insofar as they still have an external form and substance), animal life or interiority is limited to a group of beings who meet specific criteria. The real problem arises when one tries to determine these criteria: unless one believes in a divine authority or science as the source of ultimate truth, any determination of the line that separates one mode of life from another will inevitably depend on political decision. Thus, as Esposito warns us, the role of modern (bio) politics is no longer "to define the relationship between human beings as much as to identify the precise point at which the frontier is located between what is human and what . . . is other than human" (*TP*, 24). If we transpose this logic onto the relationship between various human groups with non-human beings and things in their environment, we will see, as Descola and a number of others have pointed out, that the line that divides the ensouled beings (humans) from the soulless ones (non-humans) is not set in stone. Among many human groups we encounter other types of ontology,[7] which challenge our modern, naturalist view of the world that claims to have a clear sense of the border between culture and nature, human and nonhuman. The modern view of the limits of sociality and community is contingent upon the belief in the primacy of the human mode of being and interiority, which goes hand in hand with various practices of de-humanization and reification of multiple entities in the environment, some of which might even belong to the human species. This means that while naturalist ontology takes as its starting point the affirmation of the shared materiality of all beings (exemplified by Bichat's and later Darwin's theories of life), it ultimately relies on the separation and elevation of the human above this continuity. However, this very elevation is based on the attribution to the "human" of a unique interiority, soul or, in Esposito's terms, of personhood. Thus modern naturalism, which has nurtured the exclusionary *dispositif* of the person, is paradoxical insofar as it incorporates two extremes: it affirms shared material existence of all beings and, at the same time, denies a great part of these beings any qualitative life, forcing them into the precarious existence of animals and things. Descola points out this paradox of modern naturalism and suggests that it regards the non-human, especially an animal, "either as the lowest common denominator of a universal image of humanity or else as the perfect counterexample that makes it possible to define the specific nature of that humanity."[8]

The "natural" element of life that was extracted from a living being had to be properly labeled and mastered, just as "nature" was tamed and became "dumb, odor-free, and intangible . . . all that remained was a ventriloquist's dummy, of which man could make himself, as it were, the lord and master."[9] The degree of this mastery over the natural or the inhuman part by the rational part of self, as Esposito shows, was precisely what defined modern personhood. As a result, personalization has been always "de-animalization" through which *homo* ("the human as a natural being") attained its "true" humanity. Shared "organic" life was thus suppressed and turned into a thing that can be possessed and mastered by "the personal self."[10]

Esposito suggests that after the end of the World War II we witness a return to the concept of the person as the rational mastery over the animal, the dominance of which was previously overturned by Bichat and his followers in the human sciences, as well as by Nazi ideology. In a way, the paradox of modern naturalism has been exhibited historically through the shift from *Nazi biopolitics*, which collapses the personal self into the bare biological and racial given, toward *liberal biopolitics*, which aims to restore to the human its decision-making power, rational will and to distance it from being just a body. The priority was given back to the personal over the impersonal life in attempt to bridge the gap between human being and citizen and to allow for the possibility of human self-determination and rights (*TP*, 72). However, the aspirations of the 1948 Universal Declaration of Human Rights, which cemented the priority of the person over the impersonal, did not materialize; insofar as today, over fifty years after its creation, we witness more cases of abuse of these "human rights" than of their fulfillment.

Esposito argues that this failure of the modern regime of human rights is not due to the lack of its implementation. The failure is rather inherent in the concept of the person that supports this regime: the mere positioning of the personal over the animal does not prevent the *dispositif* of the person from producing its destructive effects, because a shift in hierarchy does not overcome the problematic separation between two parts of the living being, which Bichat identified as organic and animal. Liberal biopolitics does not overcome the division between "culture" and "nature"; on the contrary, it is conditioned by their very separation. Esposito shows how the subjection of nature to human artifice has in fact a long history and can be traced back to ancient Roman legal and Christian theological matrices. Without going into detail, what is important about both of them is that they define person as that "which in a human being is other than and beyond body" and corresponds "to the irreducible difference that separates the living being from itself" (*TP*, 76). There are no persons in "nature"; they only come into being through a secondary operation of artificial aggregation through "culture." Esposito's analysis of Hobbes's theory of sovereignty is exemplary in this regard.

Hobbes establishes "the logical and semantic prevalence of the artificial person over the natural one," effectively inaugurating modern naturalist ontology. According to Hobbes, there are no persons in the state of nature, implying that it is inert and incapable of anything except war. In the state of nature there exist only "natural persons," coinciding "with his or her living (and, before long, dying) being. In other words, there is no such thing as the transcendence of self, which is the necessary condition of personality." Interestingly, Hobbes suggests that even inanimate things (e.g., a church, a hospital, a bridge) can be "personated," however, as in the case of human beings, these "feigned" persons cannot attain their status "before there be some state of civil government." As a result, the sovereign, established through a transition from the state of nature to the commonwealth, is "the agent of personalization," who has the ability to transform "things" into new legal persons, and, by the very same logic, of "de-personalization," who can thrust people into the realm of things (*TP*, 84–86). The sovereign is the gate-keeper of personhood and is in charge of the status of all the entities under his "protection": no being or thing crosses the threshold of personhood (in either direction) without the permission of the sovereign law. Our contemporary society is subject to this logic inasmuch as it was during Hobbes's time, and we can see it operating, for example, in debates on abortion, euthanasia, animal rights, and ecology, all of which hinge on the question of the *legal* attribution of personhood.

In sum, the modern *dispositif* of the person regulates the relationship between and the status of four terms—human and animal, person and thing. While the oscillation between human and animal points at the division within a living being and, subsequently, between living beings, the oscillation between person and thing relates to the status of these living beings in the legal order of society. The terms of these dichotomies can produce different combinations, for example, humans reduced to the status of things and animals or inanimate entities legally declared as persons.[11] Esposito notes that the most important and the most tragic outcome of the operation of this system of coordinates is that personalization of one entity relies on the de-personalization of another; in other words, the flip-side of the process of becoming-person is reification of the impersonal. The essential function of the *dispositif* of the person is "the summed separation, operated inside the human being, between a natural, merely biological corporeal [impersonal] element and another [personal] element, which is transcendental and construed from time to time as the site of legal, rational, and moral imputation" (*TP*, 97). It is the ongoing separation between the mask (the person) and the face (the impersonal) and the suppression of the latter by the former that defines this *dispositif*. The problem arises once the priority of the human person is made safe, as has been the case since the end of the WWII, and the fate of the impersonal is disregarded. As a result, "it does not matter what happens to the face on which it rests and even less to the faces that do not own masks; to those who still aren't persons, or who are no longer persons, or to those who were never declared to be persons."[12] Modern personhood is thus the threshold beyond which something generically living can be considered qualitatively significant and, as such, it "marks the final difference between what must live and what can be legitimately cast to death" (TP, 13). *Bio*political care, then, is restricted to care for personal life, while the impersonal becomes the object of *thanato*politics.

The Power of the Impersonal: Toward the *Dispositif* of the Living Person

In order to counter the effects of the mechanism of separation and exclusion characteristic of the *dispositif* of the person, Esposito turns toward the thought of the impersonal, the outlines of which are implicitly present in contemporary art, post-Freudian psychoanalysis and some works of twentieth-century philosophy. It is the latter that draw Esposito's attention and based on which he builds something like a genealogy of the impersonal. Esposito suggests that the thought of the impersonal or the conceptual work on the "third person," rather than merely destroying the person, creates "an opening to a set of forces that push it [the *dispositif* of the person] beyond its logical, and even grammatical boundaries" (*TP*, 14). This work of opening suggests that the impersonal is not a mere negation of the personal, but is that "which, from within the person, blocks the mechanism of distinction and separation with respect to those who are not persons."[13] It is located at "the confines of the personal; on the lines of resistance . . . which cut through its territory, thus preventing . . . the functioning of its exclusionary dispositif" (*TP*, 14). Esposito notably speaks of the impersonal in active terms (it blocks, prevents, resists, cuts through, etc.), pointing out its productivity not only in deconstructing and opposing the personal regime but also in performing a positive work of opening "in favour of a different *regime* of meaning" (*TP*, 16; emphasis added). Is it possible then that the impersonal is another *dispositif* that emerges out of the fissures and cracks of the personal? Is it not the surplus of the old apparatus? Arguably, since impersonal resistance is located at the confines of the personal and points toward a different regime of meaning, it is possible to see it as a different form of production of subjectivity that is no longer located within the confines of the personal but opens, as Esposito puts it, "to what has never been before." Deleuze writes, in this respect: "One might wonder whether lines of subjectivation are the extreme edge of an apparatus and whether they trace the passage from one apparatus to another: in this sense, they would prepare 'lines of fracture.' "[14] We can interpret the thought of the impersonal, implicit in modern philosophy dominated by the concerns of personhood, as preparing such a "line of fracture" that marks a passage from the *dispositif* of the person toward a new *dispositif* of the *living* person. If we make use of Deleuze's terminology, the impersonal is a name for a group of "lines of actualization or creativity" that counter the "lines of stratification or sedimentation"[15] within the apparatus of the person. Thus, the thought of the impersonal, in its multifaceted appearance, is not a negation of the person but takes shape as "a contrasting complimentarity . . . as a *fold* inside the larger figure it is intended to oppose" (*TP*, 8; emphasis added). It is my suggestion that this fold produces a different interiority or, rather, a different regime of distinction between the inside and the outside, which I would like to call *the dispositif of the living person*. The fracture within the personal opens not onto some chaotic attribution of properties of personhood to everything around us, but onto a more specific *regime of inclusion* that no longer operates through the mechanisms of separation and exclusion but through *folding* and *contamination.* As Timothy Campbell puts it, the impersonal weakens "the various thresholds of personhood premised on the division between human and animal, organic and inorganic" and "short-circuits self-defences [of the person] by bringing the outside in."[16]

In his search for a new *dispositif,* Esposito raises the question of "how to make the impersonal not only a power for deconstructing the ancient—and new—dispositif of the person, but also the form, or, better still, the content of a practice that alters existence" (TP, 17). He asks how the impersonal lines of fracture can shift the existing *dispositif* toward a new regime of productivity, and it is in the works of Foucault and Deleuze that he finds an indication of such a passage from the *thought* to the *practice* of the impersonal. In both cases, it is the notion of "life" that comes to the fore as the field of operation of the impersonal, but the opening it produces in this respect is twofold. First, Foucault's notion of resistance, which invariably inhabits any form of power relations, including those that seize life as their object, lays the ground for "an affirmative biopolitics." Second, Deleuze's notion of "becoming-animal" indicates more clearly a transition from the naturalist ontology of personhood, based on the division along the line of culture or artifice and nature, toward a different kind of ontology that conditions the emergence of the *dispositif* of the living person.

Esposito suggests that in his later works Foucault elaborates his notion of "the unthought" (introduced in *The Order of Things*) in terms of "life." From this perspective, the outside (i.e., the non-personal within the person that Esposito is after) is "life itself": it is what we are—"beyond or before our person—without ever being able to become masters of it," it is what "traverses us and troubles us, to the point of turning over into its opposite" (*TP*, 137). The outside is inside; the non-personal inhabits the person as its own life. Importantly, this "life" in modernity becomes "the ultimate and primary ground of battle . . . a battle in which the most extreme risk is united with the most unprecedented of opportunities" (*TP*, 138). Life is both the primary object of (bio)political government and the ground of resistance to biopower. Foucault affirms that life is never fully captured by the mechanisms of power but "constantly escapes them."[17] In Esposito's words, life "is the space that power lays siege to without ever managing to occupy it fully, even generating continuously new forms of resistance." Such an inexhaustibility of living resistances implies that biopolitical government and control over life are open to their opposite, to an opportunity of life taking charge of itself—"an affirmative biopolitics" (*TP*, 18). Life, "distinct from the subjectivity of the person as that which both underlies and overturns it into its material exteriority, constitutes the object of biopolitics, but also the locus that most opposes it." Most importantly, this Foucaultian line of reasoning enables Esposito to affirm that "the person is not to be conceived of as the only form within which life is destined to flow" (*TP*, 139, 140). Life is simultaneously within and exterior to the person; it is nameless and impersonal and thus clings only to itself.

However important Foucault's philosophy is for Esposito's project of affirmative biopolitics, his notion of the living person is most clearly articulated through a reading of Deleuze. As Esposito notes, Deleuze shifts "the entire philosophical horizon toward a theory of the pre-individual, impersonal event" (TP, 142), one of which is "*a* life"—immanent, impersonal life (given that every aspect of it eludes our control) which is entirely coextensive with itself.

> Life is the tangent, the line of force, along which immanence folds back on itself, eliding any form of transcendence or any ulteriority beyond being a living substance as such. It refers neither to a rational subject not to bare material substance. But above all, if understood in its impersonal, singular dimension, life is what does not allow—what contradicts in its roots—the hierarchical division between these two entities within the separating dispositif of the person. (*TP*, 147)

As such, "life," for Esposito, sums up the entire theory of the impersonal. He identifies in Deleuze's thought several specific lines of deconstruction of the category of the person, among which are the concepts of "virtuality," "individuation," and "becoming-animal." It is the latter, Esposito notes, that "opens thought on the impersonal onto a perspective whose significance as a whole remains to be understood" (*TP*, 19). It is my suggestion that the figure of becoming-animal outlines *the dispositif of the living person*, insofar as it deconstructs the person through an embrace of "a logic that privileges multiplicity and contamination over identity and discrimination" (*TP*, 145). While modern processes of personalization have been correlated with de-animalization, the impersonal experience of becoming-animal vindicates "animality as our most intimate nature," but not in a limited sense of recognition of the animal "ancestry" of the human species or of the importance of the so-called animal needs of human beings. Rather, following Deleuze, Esposito argues that

> [t]he animal—in the human, of the human—means above all multiplicity, plurality, assemblage with what surrounds us and with what always dwells inside us [. . .] But this also means plurivocity, metamorphosis, contamination [. . .] In opposition to purity, against its supposed immune effects, the becoming-animal becomes "propagation by epidemic, by contagion [. . .]" [I]t brings into relationship completely heterogeneous terms—like a human being, an animal, and a micro-organism; but even a tree, a season, and an atmosphere: because what matters in the becoming-animal, even before its relationship with the animal, is especially the becoming of a life that only individuates itself by breaking the chains and prohibitions, the barriers and boundaries, that the human has etched within it. (*TP*, 150)

The experience of becoming-animal is not reduced to the human-animal relationship but indicates the relationship between multiple, heterogeneous elements in an environment, which might result in composition of specific entities—living persons—whose interiorities are not predetermined but remain permeable and open to re-composition through an ongoing relationship with exteriority. At the limit, the living person does not have interiority in the sense of a fixed status (i.e., personhood with specific attributes), but only as a tentative assemblage, since the living subject is always outside, it is potentially everywhere, including all form of life's becoming. This kind of interiority, where inside is always outside, is "animistic."[18] Importantly, animism here does not merely signify a type of ontology that can be attributed to specific human groups, but is also a name of the "repressed unconscious" of modern naturalism.[19] It is my contention that Esposito develops his *dispositif* of the living person within the field of this unconscious of modernity, of which contemporary philosophy is becoming more conscious, in particular through contamination by indigenous thought. In order to develop the notion of the living person further, without the threat of sliding back to the premises and divisions of naturalism, we might benefit from directly engaging with non-Western conceptions of personhood that rely on plurivocity, metamorphosis, and contamination in the process of composition of personal unity and interiority.

Animistic Personhood

It would be overly ambitious, if not altogether impossible, to present here a detailed overview of the wide literature dedicated to the study of animism. I will only offer a brief introduction to the ongoing debates on animism in philosophy, political theory, and anthropology, as well as sketch the idea of "animistic personhood" in order to open a conversation between Esposito's notion of the living person and indigenous philosophies of personhood. Such a conversation can be helpful in furthering our understanding of what kind of social practices the *dispositif* of the living person might produce.

The term "animism," introduced into modern discourse by a German physician and chemist Georg Stahl in 1708, was adopted and popularized by the founder of anthropology, Sir Edward Burnett Tylor, who in his 1871 book, *Primitive Culture*, defined it as a primitive, erroneous religious category describing the belief in souls and spirits. It is

> an idea of pervading life and will in nature far outside modern limits, a belief in personal souls animating even what we call inanimate bodies, a theory of transmigration of souls as well in life as after death, a sense of crowds of spiritual beings, sometimes flitting through the air, but sometimes also inhabiting trees and rocks and waterfalls, and so lending their own personality to such material objects.[20]

In Tylor's evolutionary theory of religion, animism is an early stage of development attributed to "savages and barbarians": "a mythopoetic mode of discourse that explains life and events to those not yet fully acculturated to the practice of rationalist science," those who rely not on objective scientific observation but on the senses supported by dreams and visions.[21] For him, animism is the opposite of civilization and adulthood, and can be found only among "savages" and children. It is a stage of cultural development and not a legitimate ontology; a misperception of reality rather than an elaborate epistemology. Tylor's interpretation of animism has been persistent in modern anthropology that has contributed to the justifications of Western colonial practices of "education," often resulting in the destruction of indigenous groups.

"Old animism," i.e., the Tylorian-inspired understanding of animism as a primitive stage of cultural evolution, has been challenged within past three decades by so-called "new animism" that no longer asserts animism's erroneous nature and recognizes it as a legitimate, rich, and highly developed ontology that presents an alternative to modernity.[22] According to Philippe Descola, one of the prominent proponents of new animism, animistic ontology is the polar opposite of naturalism. The latter, in ordering the relationships between humans and non-humans or between culture and nature, recognizes a continuity of their physicalities, i.e., their material commonality, and a discontinuity in their interiorities, allowing only humans to be endowed with a proper "soul" or consciousness. Animism, on the contrary, assumes a continuity of interiorities and a bodily discontinuity between humans and non-humans.[23] In this ontology, consciousness, intentionality, will, and agency are not limited to humanity and can be encountered in a variety of entities in an environment. The Western distinction between human culture and non-human nature does not hold here, since culture and sociality are distributed among the heterogeneous beings. It is an ontology that postulates the social character of relations between "person" and "thing," "human" and "animal." New animism, then, is distinctively non-dualistic and "reveals itself as the recognition of the universal admixture of subjects and objects, humans and non-humans against modern *hubris*, the primitive and post-modern 'hybrids.'"[24] It relies on relational "technologies" in ordering the *socius*: identities and positions are formed through ongoing relationships and internalization of prior externalities; they are fluid and resist fixation and essentialization.

Importantly, as, for instance, Bruno Latour suggests, animism is more than just the opposite of modernity; it actually operates in modern societies despite their dominant ontological (naturalist) constitution, insofar as these societies continue to animate the world around them and to engage in relationships with non-humans, even if these non-humans are not endowed with some sort of eternal and immaterial souls, but only possess material agency.[25] I believe that Esposito's notion of the living person is moving in this direction by tapping into the animistic unconscious of modern societies and leading them toward the recognition of multiple relationships with those who were long considered non-persons. This animistic opening of Western personhood paves a path to an enriched conception of sociality, based on "a staggeringly ecumenical relationality to the world,"[26] and to "seeking better forms of personhood in relationships."[27]

As noted above, one of the particular alternatives that animistic ontology offers to modern naturalism is a non-dualistic conception of personhood. In Graham Harvey's words, "[a]nimists are people who recognize that the world is full of persons, only some of whom are human, and that life is always lived in relationship with others."[28] Animistic personhood is often referred to as "relational"[29] or "composite," pointing out that the person is always in the process of becoming, without preexisting or predetermined boundaries between inside and outside. The relationship between these "domains" occurs not through the mechanism of immunitary sanitation, to use another of Esposito's terms,[30] which prevents the outside from invading the inside, but through their constant cross-contamination and exchange. Like the living person, the animistic person is a composite unity of external and internal, *bios* and *zoé*; in other words, it disregards the naturalist version of nature/culture duality that has been enabling the distinction between body and spirit, natural life (*zoé*) and its personalization (*bios*). From this perspective, contrary to Hobbes's indication (representative of the common attitude of Western modernity) that there are no persons in the state of nature, animist ontology allows for personhood to emerge or to be encountered everywhere, including in the domain that Western mind perceives as "nature." "Person" is an overarching category and not a mere subcategory of "human": there are "human persons," "animal persons," "wind persons," "stone persons," etc.[31] All of them are or may become persons but none is *definitive* of personhood.

Importantly, not everything is a person and not every member of a species or a category of beings is automatically a person merely due to its membership. For instance, as the Ojibwa would tell you, not all stones are alive and are persons "but some are."[32] Persons are distinguished from non-persons through their relationships with others; they are "willful beings who gain meaning and power from their interactions," they are persons because they "relate and communicate."[33] In animist ontology, *personhood does not preexist relations*, that is, we cannot decide in advance of a social encounter whether something is a person. As Nurit Bird-David puts it in her famous study of animistic practices among the hunter-gatherer people of Nayaka, "[w]e do not first personify other entities and then socialize with them but personify them *as*, *when*, and *because* we socialize with them. Recognizing a 'conversation' with a counter-being—which amounts to accepting it into fellowship rather than recognizing a common essence—makes that being a self in relation with ourselves."[34] Inasmuch as no one is born a complete person but requires continuous relational growth and initiation to keep them personal, no one is born an "animist" but becomes one as he or she *learns* to recognize personhood in multiple entities in the immediate environment and to act respectfully toward them, and so the role of elders in many indigenous communities lies in helping the younger generation to become increasingly animistic.[35] It seems that in Western modernity philosophers and scientists have assumed the role of such elders and, arguably, Esposito's thought of the impersonal can be interpreted as an attempt at an animistic attunement of (post)-modern humanity by means of philosophy.

Contrary to the naturalist conception of the person, animistic personhood resists legal formalization insofar as it is not a universal but an *enacted* category that has to be reproduced through actual encounters in local environments, and here the sovereign can no longer be the gatekeeper of personhood. Person is always in the process of composition and decomposition, of transformation and metamorphosis. Animistic personhood does not adhere, like a mask, to a preexisting unified and natural substance, be it a face or a body with fixed limits defined by the outer layers of the skin, but "is a composite of transferable particles that form his or her personal substance."[36] As Bird-David further describes Nayaka technologies of personhood in this regard, "[t]hey *make* their personhood by producing and reproducing sharing relationships with surrounding beings, humans, and others. They do not dichotomize other beings vis-à-vis themselves but regard them, while differentiated, as nested within each other."[37] The Nayaka thus do not distinguish between the subject-persons and object-things, but "between persons in the process of relating and persons being disinterested in and unresponsive to others."[38] The disinterested quasi-nonpersons always contain the potential to reveal themselves as persons or to inhabit other persons as their composite parts. In sum, animistic persons (both willful and disinterested) are in a relationship of mutual *contamination* (i.e., nesting within each other) and not immunitary separation, as parts of one person can be transferred to another person, radically exposing one's interiority to the outside through the ever-present possibility of re-composition.

Marilyn Strathern, in her seminal study of society in Melanesia, adopts the term "dividual"[39] (the opposite of modern "individual") to refer to the non-unitary but also non-dualistic idea of personhood composed of a series of parts that originate outside of the person and that can as easily be transferred from one to another person or disposed of. She writes that

> for contextualizing Melanesians' views we shall require a vocabulary that will allow us to talk about sociality in the singular as well as the plural. Far from being regarded as unique entities, Melanesian persons are as *dividually* as they are individually conceived. They contain a generalized sociality within. Indeed, persons are frequently constructed as *the plural and composite site* of the relationships that produced them. The singular person can be imagined as a social microcosm.[40]

Strathern addresses the problem of the singular-plurality of personhood through the notion of dividual: the Melanesian person is not a unitary substance but is composed out of past and ongoing relations, substances, and actions of others. Persons are "multiply-authored," encompassing multiple constituent things and relations.[41] Their internal composition depends on a wide range of external relations, and these relations, in turn, are "condensed" into physical substances or objects, which can be given away. Personhood is composed and recomposed through, for instance, gift exchange, where personal substances mingle and are shared among the community members. Thus, the Melanesian person is not only multiple but also "partible, an entity that can dispose of parts in relation to others."[42] As Mark Mosko shows, composite and partible persons contain within them components from the whole community, as if sharing and mixing blood with all other members.[43] Personal and social unities are produced through the mechanism of cross-contamination rather than purification of blood, characteristic of modern societies, of which Nazi thanatopolitics is the "purest" example.

In sum, in animist ontologies, however undeniably varied and distinct they are, we can discern a common trait: personal unity is always multiple, whatever scale we look at, whether it is one person, one family, one clan, etc.; persons appear on different scales, but they all are similarly multiply composed.[44] Animistic personhood is singular-plural. Similarly, the notion of the living person refers to such an assemblage of multiplicities, to the "being that is both singular and plural" (*TP*, 151), the unity of which is defined through re-composition and contamination from the outside, and is produced not through a closure around the person, as in the modern *dispositif*, but through an ongoing folding and unfolding of life. The interiority of the living person does not exist as its property, because it is constantly shared and expropriated by other human and non-human living persons. In this regard, the living person is animistic insofar as it is dividual and composite, multiple, and partible. The animistic or living person then is not a closed entity but a temporary form of stabilization and shaping of relations in terms of trends or *pathways*.

Not a necessary attribute of good or qualified life, living personhood is nevertheless a sort of prosthesis that allows performance of certain roles, functions, and metamorphoses. Perhaps, this is why Esposito, paradoxically, after a lengthy deconstruction of the modern *dispositif* of the person, in the end retains the notion of the person, however alternatively constructed. The idea of personhood as a prosthesis is old, inasmuch as it brings us back to one of the ancient Roman origins of the term—*personae*—the masks used by actors in theatrical performances. As Esposito mentions on several occasions, the separation between the face of an actor and the mask is representative of the separation between *zoé* and *bios*, natural impersonal life and the social artifice of personhood that indexes the modern *dispositif* of the person. However, masks in other "cultures" do not function only as they did in ancient Greek and Roman theatres. While putting on a mask or becoming a person in Western society relies on negation or suppression of animality underneath it, interactions with a mask in many animistic communities during rituals and ceremonies rely on the opposite movement of becoming-animal. For example, the masks used by many Amerindian indigenous peoples are not human but animal masks that, as Eduardo Viveiros de Castro explains, are meant

> not so much to conceal a human essence beneath an animal appearance, but rather to *activate the powers* of a different body. The animal clothes that shamans use to travel the cosmos are not fantasies but *instruments*: they are akin to diving equipment, or space suits, and not to carnival masks. The intention when donning a wet suit is to be able to function like a fish, to breathe underwater, not to conceal oneself under a strange covering.[45]

The mask here is a prosthesis of transformation and metamorphosis and not of concealment and negation of the impersonal face underneath it; the animist mask-*persona* enables one to walk particular paths, to engage in and maintain particular patterns of behavior, to develop previously inaccessible powers and capacities, thus constantly producing and extending the field of encounters and relationships.

The unique transformation masks used by the indigenous peoples of the Pacific Northwest Coast and Alaska are a particularly good example of the productivity of the animistic *personae*.[46] These large masks, with movable parts that can be opened and closed, usually depict on the outside an animal visage and when opened reveal an image of a human face. The metamorphosis occurring during ceremonial dances references the mythological time when all living beings were unified and animals could take on human form, just as easily as humans could become animals. Most importantly, the internal layering of the mask implies the composite essence of personhood and reveals that underneath a mask one does not encounter a natural entity (a face) but just another mask. The mask cannot be separated from the face because there are only masks underneath masks. From this perspective, the living person can be seen as a mask that empowers a life, in an almost shamanic way, to overcome the barriers that the human has etched within it and to develop new powers. The notion of the person, instead of remaining a mechanism of destruction, transforms into an *instrument* of a life's becoming and empowerment—the living person. Although in *Third Person* Esposito does not connect his idea of the singular-plurality of the living person with animistic conceptions and apparatuses of personhood, I think it is worth to seriously consider non-Western, indigenous perspectives on personhood at the time when the living person is still, as Esposito puts it, only a "figure that has yet to be fathomed."[47] The *dispositif* of the living person might not yet be productive of real effects in Western societies marked by the domination of naturalism, but it has been operational in animistic societies, and so we can rely on the knowledge and wisdom of previously marginalized indigenous philosophies in reconfiguring our thought and practices in terms of the *dispositif* of the living person.

One of the questions that remains regarding Esposito's notion of the living person is what effects it might produce for theory and practice. First, we need to see more clearly how living personhood operates in practice and what "forms" it is capable of assuming *beyond legal formalization*. This is an important task if we want to show that the living person can do more than offer itself, as Alberto Moreiras points out in the present volume, as a mere "ideological banner to ward off the most egregious abuses against disposable human life." Is there more to the living person than an idea? It is my contention that turning to the experiences of animistic societies not only helps us elaborate the *notion* of the living person (which is the primary object of the present chapter) but also offers examples of the real operation of its *dispositif*. In this regard, we need to further examine the forms that the living person assumes in indigenous sociopolitical and ethical practices.

Second, we need to assess the consequences that the opening, offered by the *dispositif* of the living person, produces for political theory. Political theory is often said to begin "with (and as) an active purification of human society from the material world, in which the idea of humankind's removal from a state of nature marks the threshold of civilization and the possibility of political order."[48] Polis has been long seen as a place of public debate, from which "nature" is excluded or at most "represented" through human needs and desires. What happens to political theory when the "new" public includes no longer merely *legal* but also *living* persons, who bring into relationship a human being, an animal, a micro-organism, a tree, a season, and an atmosphere? Importantly, we cannot stop halfway and allow only scientifically probable or provable existents to be accepted as participants in the composition of the living personhood and thus as a part of the new public. We have to go as far as to recognize that "speculative" nonhumans play a part in the relationships constitutive of the living personhood. If we do not do that, then we doom political theory to become a tool of censorship and a mechanism of exclusion based on the primacy of the scientific method, bringing us back to naturalism. However, if we do radically open public space and political theory to the living person, then are we not, as Isabelle Stengers asks, opening Pandora's box?[49] The promise contained in the *dispositif* of the living person, at first glance, presents two alternatives: first, a radical opening of the political field to the nonhuman that results in the destruction of any specificity of politics and thus leads to the disappearance of political theory in favor of now totally politicized ontology; second, the institution of a new rigid limitation of personhood based on the censorship of the scientific method, resulting in technocratic totalitarianism. However, the combination of these two alternatives produces a different outcome. If we accept that the *dispositif* of the living person (as any other apparatus), despite the opening requires a closure or can function only insofar as it is capable of producing a threshold of inclusion, we need to contemplate the nature of this threshold. It is my contention that between a pure opening and a new closure, between erasure and the redrawing of the limit lies a *speculative* threshold, susceptible to constant redefinition based on the affirmation of the ever growing potentialities. From this point of view, the threshold that defines the *dispositif* of the living person contributes to *speculative political theory,* which essentially eschews a clear and stable determination of the interiority and the exteriority of personhood. The speculative method indeed opens our thinking to "what has never been before," since there is no certainty about what can in the future contribute to the interiority of the living personhood and thus constitute a novel subject of politics. There is no knowing what politics, defined by the speculative threshold, may become, and whether, in its radical opening, it might not cease to exist. I believe that this line of thinking is the most productive and promising today insofar as it challenges us to reconsider the founding premises of Western tradition of political philosophy and forces us to acknowledge the real possibility of political finitude. What needs to be done in this regard is a further philosophical examination of politics that has become conscious of its limit, the task that bridges Esposito's thought of the living person with the project of various speculative realist philosophies.[50]

Alternatively, it is possible to argue that the living person and the challenge it presents for political theory and praxis is not so much an opening, as Esposito claims, to "what has never been before," but more a recognition that nonhumans were never fully cast out of the political fold. It may be that the relationships composing the living person have been always at play and have been merely marginalized through the very operation of political theory built on the premise of human exceptionalism. Challenging this premise, then, would be the first step to unveiling the transformative power of the living person—the new subject of *posthumanist political theory*. A lot of work has been already done in this regard, contributing to the project of new materialist politics,[51] and Esposito's work on the living person needs to be further situated within, compared to and integrated with the efforts of other posthumanist thinkers.

Notes

1. Esposito overall identifies "three semantic areas" of the impersonal challenge to the domination of the personal in modern philosophy. These are "justice" (represented by the philosophy of Simone Weil), "writing" (represented by the philosophy of Maurice Blanchot), and "life" (represented by the philosophy of Michel Foucault and Gilles Deleuze). See Roberto Esposito, "For a Philosophy of the Impersonal," *CR: The New Centennial Review* 10, no. 2 (2010): 121–34.

2. Eduardo Viveiros de Castro in Angela Melitopoulos and Maurizio Lazzarato, "Assemblages: Félix Guattari and Machinic Animism," *E-Flux* 36 (July 2012), http://www.e-flux.com/journal/assemblages-felix-guattari-and-machinic-animism/.

3. For further discussion of the notion of *dispositif* see, for example, Gilles Deleuze, "What Is Dispositif?," in *Two Regimes of Madness: Texts and Interviews 1975–1995*, ed. David Lapoujade (Los Angeles, CA; Cambridge, MA: Semiotext(e),

2006), 338–48; Giorgio Agamben, *What Is an Apparatus?* (Stanford, CA: Stanford University Press, 2009).

4. Roberto Esposito, "The Dispositif of the Person," *Law, Culture and the Humanities* 8, no. 1 (2012): 22.

5. Xavier Bichat, *Physiological Researches on Life and Death* (London: Longman, 1815), 21.

6. Philippe Descola, *Beyond Nature and Culture* (Chicago: University of Chicago Press, 2013), 116, 121.

7. Descola identifies four major types of ontology: "totemism," "analogism," "animism," and "naturalism" (Descola, *Beyond Nature and Culture*, 121).

8. Descola, *Beyond Nature and Culture*, 178.

9. Ibid., 61.

10. Timothy Campbell, "'Enough of a Self': Esposito's Impersonal Biopolitics," *Law, Culture and the Humanities* 8, no. 1 (2012): 32.

11. There are a number of examples of these combinations. Esposito discusses the status of slaves in ancient Rome as an example of reification of human beings. We can think of more recent examples of the opposite process, where non-humans are declared as *legal* persons; for instance, India granted dolphins and whales the status of "non-human persons" in 2013, and Bolivia in 2010 granted "Mother Earth" rights akin to human rights, in particular, the right to life (*Ley de Derechos de la Madre Tierra* (Law of the Rights of Mother Earth)).

12. Esposito, "The Dispositif of the Person," 30.

13. Esposito, "For a Philosophy of the Impersonal," 130.

14. Deleuze, "What Is Dispositif?," 341.

15. Ibid. 347.

16. Campbell, "'Enough of a Self,'" 44, 46.

17. Michel Foucault, *The History of Sexuality: An Introduction. Vol. 1* (New York: Vintage Books, 1990), 143.

18. Viveiros de Castro in Melitopoulos and Lazzarato, "Assemblages."

19. Cf. Dietrich Karner in Anselm Franke and Sabine Folie, eds., *Animismus: Moderne Hinter Den Spiegeln* [*Animism: Modernity through the Looking Glass*] (Köln: Buchhandlung Walther König, 2011), 7. Karner notes that animism has been retained as a legitimate perspective in modernity in the forms of infantile desires, fantasies, mental illness, and art. Interestingly, Esposito notes that traces of the thought of the impersonal can be located, beyond philosophy, in psychoanalysis and in art, implying, perhaps, a conclusion that the impersonal is another name for animism as the repressed unconscious of modernity.

20. Edward B. Tylor, *Primitive Culture: Researches into the Development of Mythology, Philosophy, Religion, Art, and Custom*, vol. 1 (London: J. Murray, 1871), 260.

21. Graham Harvey, *Animism: Respecting the Living World* (New York: Columbia University Press, 2013), 6, 8.

22. Among the names commonly associated with "new animism" are Irving Hallowell, Nurit Bird-David, Tim Ingold, Philippe Descola, Eduardo Viveiros de Castro, Rane Willerslev, Bruno Latour, and Graham Harvey. For a thorough discussion of the historical emergence of the notion of animism and its contemporary theoretical transformation from derogatory to critical term, see Harvey, *Animism*, 3–29.

23. Descola, *Beyond Nature and Culture,* 121.

24. Eduardo Viveiros de Castro, "Cosmological Deixis and Amerindian Perspectivism," *The Journal of the Royal Anthropological Institute* 4, no. 3 (1998): 475.

25. In this regard, see, for example, Latour's discussion of "actants" in Bruno Latour, *Politics of Nature* (Cambridge, MA: Harvard University Press, 2004), 70–77. Jane Bennett's work also is valuable in this regard. Through an adaptation of Latour's concept of "actant," she discusses multiple ways in which matter is animated and acts in relation to humans in everyday interactions (Jane Bennett, *Vibrant Matter: A Political Ecology of Things* (Durham, NC: Duke University Press, 2010).

26. Timothy Campbell, "'Foucault Was Not a Person': Idolatry and the Impersonal in Roberto Esposito's *Third Person*," *CR: The New Centennial Review* 10, no. 2 (2010): 149.

27. Harvey, *Animism*, 16.

28. Ibid., xi.

29. Nurit Bird-David in her revisitation of animism calls it, more specifically, a "relational epistemology" (Nurit Bird-David, "'Animism' Revisited: Personhood, Environment, and Relational Epistemology," *Current Anthropology* 40, no. S1 (1999): S67–S91).

30. Cf. Roberto Esposito, "The Immunization Paradigm," *Diacritics* 36, no. 2 (2006): 23–48; Roberto Esposito, *Immunitas: The Protection and Negation of Life* (Cambridge, UK; Malden, MA: Polity Press, 2011).

31. Cf. Bird-David's commentary on Irving Hallowell's study of the Ojibwa (Bird-David, "'Animism' Revisited," S71).

32. Irving Hallowell, in his famous study of the Ojibwa ontology, recounts an anecdote: "Since stones are grammatically animate, I once asked an old man: Are all the stones we see about us here alive? He reflected a long while and then replied, 'No! But *some* are.' This qualified answer made a lasting impression on me. And it is thoroughly consistent with other data that indicate that the Ojibwa are not animists in the sense that they dogmatically attribute living souls to inanimate objects such as stones. . . . It does not involve a consciously formulated theory about the nature of stones. It leaves a door open that our orientation on dogmatic grounds keeps shut tight. Whereas we should never expect a stone to manifest animate properties of any kind under any circumstances, the Ojibwa recognize, *a priori*, potentialities for animation in certain classes of objects under certain circumstances" (Alfred Irving Hallowell, "Ojibwa Ontology, Behavior and World View," in *Culture in History: Essays in Honor of Paul Radin*, ed. Stanley Diamond (New York: Columbia University Press, 1960), 24).

33. Harvey, *Animism,* 18, 100.

34. Bird-David, " 'Animism' Revisited," S78.

35. Cf. Harvey, *Animism,* 99–114.

36. Mattison Mines in Bird-David, " 'Animism' Revisited," S72.

37. Bird-David, " 'Animism' Revisited," S73.

38. Harvey, *Animism,* 101.

39. Marilyn Strathern, *The Gender of the Gift: Problems with Women and Problems with Society in Melanesia* (Berkeley: University of California Press, 1988). While popularized by Strathern, the term "dividual" was coined by McKim Marriot in his study of Indian personhood and caste (McKim Marriot, "Hindu Transactions: Diversity without Dualism," in *Transaction and Meaning: Directions in the Anthropology of Exchange and Symbolic Behavior,* ed. Bruce Kapferer (Philadelphia, PA: Institute for the Study of Human Issues, 1976), 109–37).

40. Strathern, *The Gender of the Gift,* 13; emphasis added.

41. Chris Fowler, *The Archaeology of Personhood: An Anthropological Approach* (London; New York: Routledge, 2004), 26.

42. Strathern, *The Gender of the Gift,* 185.

43. Mark S. Mosko, "Motherless Sons: 'Divine Kings' and 'Partible Persons' in Melanesia and Polynesia," *Man* 27, no. 4 (1992): 697–717.

44. Roy Wagner calls this type of personhood, similarly reproduced on different scales, "fractal" (Roy Wagner, "The Fractal Person," in *Big Men and Great Men: Personifications of Power in Melanesia,* ed. Maurice Godelier and Marilyn Strathern (Cambridge, UK: Cambridge University Press, 2008), 159–73).

45. Viveiros de Castro, "Cosmological Deixis," 482; emphasis added.

46. On the Native American mask tradition of the Pacific Northwest Coast see, for example, Peter L. Macnair, Robert Joseph, and Bruce Grenville, eds., *Down from the Shimmering Sky: Masks of the Northwest Coast* (Vancouver; Seattle: Douglas & McIntyre; University of Washington Press; Vancouver Art Gallery, 1998); Claude Lévi-Strauss, *The Way of the Masks* (Vancouver: Douglas & McIntyre, 1982).

47. However, in his recent book *Persons and Things,* Esposito does engage with indigenous thought and practices. See Roberto Esposito, *Persons and Things: From the Body's Point of View* (Cambridge, UK; Malden, MA: Polity Press, 2015).

48. Bruce Braun and Sarah Whatmore, "The Stuff of Politics: An Introduction," in *Political Matter: Technoscience, Democracy, and Public Life,* ed. Bruce Braun and Sarah Whatmore (Minneapolis: University of Minnesota Press, 2010), xiv.

49. Isabelle Stengers, "Including Nonhumans in Political Theory: Opening Pandora's Box?" in *Political Matter: Technoscience, Democracy, and Public Life,* ed. Bruce Braun and Sarah Whatmore (Minneapolis: University of Minnesota Press, 2010), 3–31.

50. For an introductory and more specialized discussion of speculative realism, speculative materialism, and object-oriented philosophy see, for example,

Ray Brassier, Iain Hamilton Grant, Graham Harman, and Quentin Meillassoux, "Speculative Realism," *Collapse* 3 (2007): 306–449; Levi Bryant, Nick Srnicek, and Graham Harman, eds., *The Speculative Turn: Continental Materialism and Realism* (Melbourne, Victoria, Australia: RE. Press, 2011); Paul J. Ennis, *Continental Realism* (Winchester, UK; Washington, DC: Zero Books, 2011); Peter Gratton, *Speculative Realism: Problems and Prospects* (London; New York: Bloomsbury Academic, 2014); Louis Morelle, "Speculative Realism: After Finitude, and Beyond?," *Speculations* III (September 2012): 241–72. In addition to a general discussion of speculative realism, Morelle's paper offers a bibliography of speculative realism, listing major primary works and secondary sources, as well as journals and blogs dedicated to this subject.

51. In this regard, see, for instance, Bennett, *Vibrant Matter*; Diana H. Coole and Samantha Frost, eds., *New Materialisms: Ontology, Agency, and Politics* (Durham, NC: Duke University Press, 2010); Elizabeth A. Grosz, *Becoming Undone: Darwinian Reflections on Life, Politics, and Art* (Durham, NC: Duke University Press, 2011); Richard Grusin, ed., *The Nonhuman Turn* (Minneapolis: University of Minnesota Press, 2015); Joshua Johnson, ed., *Dark Trajectories: Politics of the Outside* (Miami, FL: [NAME] Publications, 2013); Latour, *Politics of Nature*.

6

The Failure of the Political Concept of the Person?

A Foucaultian-Arendtian Response to Roberto Esposito

Antonio Calcagno

The philosopher Roberto Esposito rightly claims in his book *Third Person* that the concept of the person has been widely employed over the last two centuries to define and protect the rights of individuals in society. By defining living individuals as persons, political philosophers and practitioners have sought to safeguard the inalienable dignity and value of human life. The concept of the person has its roots in ancient Roman jurisprudence and we find mention of it in Cicero as well as the *Digest*. It is a term that is invested with different kinds of meanings, including legal, scientific, psychological, and theological ones, as it moves through Antiquity, the Middle Ages, the Renaissance, Modernity and the nineteenth and twentieth centuries. The reason why it is a significant political notion is that it defines the basic social unit of many collective political entities as well as what is owed, read "justly owed," to each and every one within a collective context: it is a basic building block of political realities.

In the nineteenth and twentieth centuries, the notion of the person was mobilized in order to argue for greater rights, freedoms, and protections for various groups in society that did not have them. In Canada, for example, women were granted rights to vote because they were deemed to be persons under the law and, therefore, entitled to certain civic privileges extended to all persons. The Minister of Justice put forward the petition to British Privy Council on behalf of the Famous 5 in 1927, asking "*Does the word 'Persons' in section 24 of the British North America Act 1867, include female persons?*"[1] The answer we know was affirmative. The Ontario reformer Agnes McPhail argued for the reform of prisons on the grounds of excessive cruelty to persons. Over the last 100 years rights extended to persons have grown and have therefore helped extend privileges, obligations, and protections to groups of Canadians that were previously ignored or even considered criminal or ill. Foucault's work on prisons and mental illness demonstrates similar kinds of shifts in Europe. This being said, the notion of person has also been abused or ignored: we see this in the case of the fight for First Nations' land rights in Canada.[2] We see how under National Socialism and Fascism, the German, Spanish, and Italian states considered many groups of peoples and minorities as non-persons. The 1948 UN Declaration of Universal Human Rights was born as an attempt to safeguard the freedom, equality, and security of the person in light of the devastation of the two World Wars.

Roberto Esposito argues that the concept of the person inevitably creates a cleavage between an actual individual life, understood outside the definitions of law, and the person, as defined by law and political practices. Giorgio Agamben argues further that because the state can exempt itself from its own laws, especially in times of crisis and attack, it can and has indeed dispensed with rights and protections normally attributed to persons under the law. In short, we find in Esposito the claim that the concept of the person inevitably ends up being manipulated by various political powers to include and exclude, classify and subject, protect and dispense with persons or non-persons as these powers see fit—all acts of Foucaultian governmentality. He subsequently argues that we should think the political concept of the person anew because it can no longer resist the machinations of politics and biology. He ultimately posits alternative concepts that he deems more viable, including what he calls the impersonal, the negative self, or the third person.

I wish here to defend the concept of the person. I argue that Esposito's argument confines personhood to legal and political discourses, thereby ignoring the impact of other discourses, including moral, psychological, and philosophical ones, which can counteract the limits that Esposito sees as detrimental for the viability of the concept. Employing the assistance of Hannah Arendt and Michel Foucault, I want to thicken the concept of political personhood by including in it a moral dimension as well as the care of the self that can overcome charges of political nonviability through the affirmation of personal identity.

Before we proceed further, I would like to make clear what I mean by personhood. I wish to confine my discussion to a biopolitical or Foucaultian paradigm.[3] I define the term in the pragmatic Foucaultian sense of a *dispositif* or apparatus. An apparatus designates an ensemble of relations "consisting of discourses, institutions, architectural forms, regulatory decisions, laws, administrative measures, scientific statements, philosophical, moral and philanthropic propositions—in short, the said as much as the unsaid. Such are the elements of the apparatus. The apparatus itself is the system of relations that can be established between these elements."[4]

A person, then, can be understood as a system of tacit and explicit relations of various discourses where one finds the term invoked. The contemporary political use of personhood, for example, has been largely informed by the atrocities of the Second World War. In political discourses, one is a person under law: she or he is extended citizenship or some legally recognized civic status that guarantees participation in society through rights, obligations, and protections. Those individuals previously excluded from or marginalized in society due to age, race, ability, gender, sexual orientation, etc., now, as persons, are recognized as having a certain right to participate fully in society. In Canada, for example, a person is defined as possessing the constitutional right to move and take up residence across the nation in order to work and earn a living; she or he also has the right to the security of her or his person.[5] For Esposito, the notion of person must be viewed from a series of power relations embodied largely in the law, politics, science, and technology.

Esposito and the Biopolitics of the Person

Esposito's argument for abandoning the concept of the person begins with a premise: there is a distinction between biological life and political life. Philosophy has classically upheld this distinction, and we see it in Aristotle's famous distinction between *zoé* and *bios*. And we see it in modern philosophy with distinctions between "man," "woman," and "citizen" or, in Locke's case, the distinction between person and human. The human belongs to the realm of nature, and humans can injure or harm one another willy-nilly, whereas a person, who is endowed with reason, may not do so willfully and negligently. The person becomes accountable, for he or she is a locus of praise, responsibility, guilt, etc., as Locke points out in his *Essay Concerning Human Understanding* (II.27.18). Esposito claims that the distinction between biological life and political life has collapsed, for political practices and security controls have become so powerful and pervasive that they can regulate, manage, and discipline the biological life upon which the concept of the person rests. Let me explain.

Esposito argues that two significant shifts in modern history have paved the way for the reduction of political life to biological life, namely, changes in the way we view sovereignty and the new biology. Modern political sovereignty has given way to management of vast populations through political and biological means. Foucault calls this governmentality or pastoral power, where the state shepherds the lives of people and not persons. Esposito remarks,

> First, [there is] the progressive transformation of the paradigm of sovereignty into that of government, given that the conditions of life of the population, its sustenance, its vital needs, begin to enter into the political objectives of power. It is at the end of the 17th century that urban, demographic, and health politics began to appear in that horizon that we can define as biopolitical. The population is no longer considered by the sovereign as something to be taken advantage of, a resource to be consumed, and it becomes a precious good that needs to be protected, a richness that it is better to preserve and develop . . .
>
> But another crucial event in this paradigmatic transformation of the preceding conceptual structures was constituted by the birth, at the beginning of the 19th century, of that discipline that is now called biology. . . . [W]ith biology the horizon of history enters into a tighter relationship with that of nature. Politics precisely situates itself at the point of contact or, often times, the point of tension, between history and nature. The human being begins to be considered as a member of a species [and not persons], just as the human species enters into contact with other living species. This determines a process of progressive de-subjectification, that is, of modification and crises of political subjectivity.[6]

The basic building block of political society that we identified as the individual person no longer functions in the way it once did, as societies have become globalized and function on scales registering huge populations never witnessed up until now.

The vast populations that constitute political states or entities have become so enormous that individual persons and the rights accrued to them become insignificant. We see this in recent discussions of national security and risk. Esposito sets these discussions of security and risk within the paradigm of immunity. In order to protect or immunize themselves, states are resorting to more comprehensive strategies of control and surveillance that require greater risk for and violence against their own citizens (*IM*, 112). The concept of the person does precious little to stop states from taking stark measures to preserve the life of the nation. The life of the state is more important than that of individual persons. We see this in Canada where the Canadian Spy Agency (only meta-data they say) has taken to carrying out greater surveillance of private individuals, all in the name of national security or the development of new tracking technologies. Care for individual persons' privacy is sacrificed for the security of the state.[7]

More recently, Esposito has argued that the relation between the concepts of the person and thing have been transformed by both the practices of the law and philosophy.[8] Though we know that the origin of the word person lies in the ancient Greek word for a face (*prosopon*) and the face masks worn in theatre, it was Roman law that gave the person a legal status. Personhood was largely determined by what one owned and what could be taken away: one had a personhood, and one was not a person. The link between personhood and ownership has been transmitted through time and has been exploited, often with noxious results, is the modern epoch, as will be discussed later in the chapter: personhood could be legally revoked and restored, by and through the law. Personhood, Esposito argues, has become a thing. But the notion of the thing, through Western metaphysics' conceptualization of it, has been so idealized that it has lost its connection to worldly reality.[9] In short, what has facilitated the vast governmentalization (through biology) of the person is its reduction to a thing and, to make things even more complex, the thingly status of personal ownership has been rendered abstract by the metaphysical idealization of the very concept of thingness.

Esposito and the Introduction of the Impersonal, Third Person, or Negative Self

If we accept Esposito's (and Foucault's) analysis that contemporary politics has taken control of life by governing, managing, classifying, regulating, surveilling, and controlling it—from birth to death—then how must we understand life, especially individual life? Has it been completely reduced to the management of our biology by politics, what Foucault calls governmentality? Esposito argues that while politics tries to govern and control life by regulating vital biological processes like birth and death, biological life itself has the potential to resist, albeit not absolutely, political interventions aimed at managing and controling it. How does life do so?

For Esposito, life is not simply a generic concept; it needs to be highly individuated and it does so through embodiment or what he calls incorporation (*IM*, 112–13). The meaning of corporeal or embodied life occupies a significant space in Esposito's work and I cannot develop it fully here, but I do wish to highlight the significant constitutive elements of his view. Corporeal individuation establishes confines or borders, and confines are exactly what acts as a line of defense against whatever threatens to take life away from itself, expels it to the outside, reverses it into its opposite (*IM*, 113). It is only in the body that life can remain what it is and even grow, reproduce, and be strengthened. Esposito describes the body as a privileged locus for the unfolding of life (*IM*, 113). The body, in addition to being the locus of life, is also the site where death becomes visible in phenomena like aging, sickness, and physical deterioration. The goal of the body is to try to prevent, as long as possible, death and sickness; it struggles against its inevitable demise and as such becomes the "battleground" or "frontline" where the struggle between life and death occurs (*IM*, 114). The body uses its own immunological process to defend itself, thereby creating the conditions that can guarantee its own survival. This is an autonomic process, and political interventions undertaken to control biological life, at least until now, face the risk of being undone by the body's own immunity system. Of course, this is not an absolute given, for the immune system also runs the risk of failing. Unlike Foucault, Esposito sees embodied biological life as possibly altering, resisting, fighting political attempts to control and manipulate it. He emphasizes this claim in *Persons and Things*, noting that embodiment reconfigures the relations between persons and things in the sense that embodiment, as it individuates and collectivizes itself can be said, following Simondon, to be "transindividual"

(*PT*, 133–34). Esposito says that the body can resist the law, because the law has not been able to form a concept of the body outside of ownership. Moreover, unlike the paradigm of ownership that typifies personhood, one *is* a body: one does not only have a body (*PT*, 108, 118).

Within the very core of the immunological paradigm, there exists a zone of indeterminacy or ambiguity: at the same time that an individuated corporeal being lives, it simultaneously dies. We all know that as we age, we die; we live while dying. The body's immune system is constantly working to rid the organism of potential life-ending threats, and does so continuously until life ceases. The third person, the impersonal or the negative self, all equivalent terms, designates that originary space or zone situated in the body's own immune system, where the body kills threats and parts of itself in order to continue living. In this space, according to Esposito, the biological differentiation between living and dying is not quite firmly delineated (*IM*, 151): one is not sure at what point something is living or dying.

If Esposito situates the negative self or impersonal within the immunological paradigm where an originary space arises between living and dying of an organism: the self can be conceived neither in a substantive sense nor in identitarian terms. He writes, "Rather than a first person, that self has become a third person: not a 'he' or 'she,' but the non person who bears both its reality and its shadow." Esposito's third person is also described as the impersonal (*TP*, 19–20). The impersonal is understood as a confluence, a crossing (*varco*) of various forces.

> . . . rather than acting as a barrier for selecting and excluding elements from the outside world, [the immune dynamic] acts as a sounding board for the presence of the world inside the self. The self is no longer a genetic constant or a pre-established repertoire, but rather a construct determined by a set of dynamic factors, compatible groupings, fortuitous encounters; nor is it a subject or an object, but rather, a principle of action. . . . It is never original, complete, intact, "made" once and for all; rather, it constantly makes itself from one minute to the next, depending on the situation and encounters that determine its development. Its boundaries do not lock it up inside a closed world; on the contrary, they create its margin, a delicate and problematic one to be sure, but still permeable in its relationship with that which, while still located outside it, from the beginning traverses it and alters it. (*IM*, 169)

Esposito's negative self certainly fits within a stream of nineteenth and twentieth century continental European philosophy, from Kierkegaard, Nietzsche, Heidegger, Deleuze, etc., that wishes to undo modern notions of the subject. His specific contribution to this debate can be seen in both his original political analysis as well as his view of selfhood localized in our immunity system. He writes,

> From an immunological standpoint, the "self" is defined only negatively, based on what it is not. This is the paradoxical conclusion, but one that is irrefutable, at least starting from the assumptions of the interpretative approach that passes from Ehrlich to the theory of clonal selection: as implicit in the evocative concept of *horror autotoxicus*, if the self recognized itself in an immune form, it would annihilate itself. The only way to survive is to be unaware of oneself. The object of the immune function, in short is never the self (except, of course, in the catastrophic case of autoimmune diseases) but rather everything that is not self. The "self" can only be immunologically expressed in the negative. (*IM*, 175)

Esposito's critique of the concept of the person (see *TP*, ch. 1) is premised on a certain reading of developments in the discourses of politics (law), metaphysics, and biology. Politics, law, metaphysics, and biology, though powerful discourses, also find themselves intersecting with other discourses, and I maintain that these other discourses add to the concept of person valuable and forceful elements that can help broaden and justify the viability of the concept. If we truly are to view the person in terms of a Foucaultian apparatus or *dispositif*, then I venture that powerful discourses in philosophy, for example, those of Hannah Arendt and Foucault, can be drawn upon to resist the restricted political views that Esposito articulates. Life not only operates at the level of biology and politics, but there are also philosophically robust discourses of moral personhood and selfhood that can resist Esposito's rejection of personhood.

Hannah Arendt and Michel Foucault: The Introduction of Competing Discourses

In her analysis of the concentration camps in *The Origins of Totalitarianism*, Hannah Arendt admits that the breakdown of the legal concept of the person, achieved through the classification of Jews and others as non-persons, was an important step that made the annihilation of millions possible. She remarks,

> The first essential step on the road to total domination is to kill the juridical person in the human. This was done, on the one hand, by putting certain categories of people outside the protection of the law and enforcing at the same time, through the instrument of denationalization, the non-totalitarian world into recognition of lawlessness; it was done, on the other, by placing the concentration camp outside the normal judicial procedure in which a definite crime entails a predictable penalty.[10]

Esposito argues a very similar position about the law and its being coopted by politics.

But Arendt also makes an important point regarding another discourse, namely, morality. In order for the Nazis to succeed, they needed to break down not only legal concepts but also ones inscribed in moral psychology. Arendt argues that the breakdown of the legal concept without a moral breakdown of the person would have given prisoners a chance to hold on to their conscience and the just conviction that they had been unjustly condemned: resistance would have been more possible. She remarks,

> Totalitarian terror achieved its most terrible triumph when it succeeded in cutting the moral person off from the individualist escape and in making the decisions of conscience absolutely questionable and equivocal. When a man is faced with the alternative of betraying and thus murdering his friends or of sending his wife and children, for whom he is in every sense responsible, to their death; when even suicide would mean the immediate murder of his own family—how is he to decide? The alternative is no longer between good and evil, but between murder and murder . . . . Through the creations of conditions under which conscience ceases to be adequate and to do good becomes utterly impossible, the consciously organized complicity of all men in the crimes of totalitarian regimes is extended to the victims and thus made really total.[11]

Arendt signals the existence of a moral person who knows how to distinguish right from wrong, and who draws value and worth from the decisions he or she makes, decisions that steer the person away from causing unjust harm, to others and to the self. From its inception, philosophy has thought about ethics, what is right and what is wrong, what is a just character and an unjust one. Arendt sees Socrates as the thinker who can distinguish between good and evil, and he shows us why it is better to choose the good and avoid the harm that ensues from evil. For Arendt, this culminates in moral personhood, which she views as critical for the good functioning of any political society. I read Arendt's later views on judgment back into the moral person. We have a lot of evidence that shows us what happens when moral personhood breaks down and is jettisoned from societies. I identify morality or ethics as an important discourse that relates to purely juridical or political discourses, understood in the Foucaultian sense, and which can have the power to stop the atrocities and violence Esposito rightly notes is antithetical for the collective flourishing of human life. Though moral personhood has no absolute force, it can still deeply condition our desire to form societies where the human good and good of all living things can flourish.

We must include in our analysis a profound sense of personhood that is not exclusively tied to politics and biology, and this is a concept of personhood that is intimately tied to personal identity or selfhood. Esposito himself denies the link between personal identity and the impersonal or the negative self. Foucault reminds us that the self is a concept that comes to be within the ancient Greek world. He reads the famous dictum "Know thyself!" as "Take care of oneself!" Later Hellenistic philosophers and theologians developed techniques of the self that could be practiced, for example, *askesis*, in order to help self and city flourish. We see here the influence of Pierre Hadot on Foucault's thinking. As he shows in volume III of the *History of Sexuality*, the early centuries of our common era, the tension between Christian and pagans over sexuality reveals an important development in the notion of the self, understood as a sexual being. Whereas early Christians viewed most sexual acts as sinful, later Greeks practiced a variety of forms of sexual acts, mindful of excess and abuse: they thought such practices contributed to the flourishing of their lives, both as individuals and as collective dwellers in the polis. Western philosophy, in its debates over the nature of personal identity, has for the most part defended the existence of the self and personal identity. Charles Taylor's magnificent work *Sources of the Self* chronicles the modern philosophical development of the notion, and shows how many of the notions developed in modern philosophy continue to influence us today, especially around property and authenticity, and so forth. What belonged to the self was what was proper to it; one owned oneself, one possessed oneself, and it was a mark of freedom to be able to do so. This freedom belonged to all selves, to all persons, to everyone, and bespeaks the unique personality of its holder. Later modern philosophy and psychology worked together to develop the notion of personal identity in terms of personality and personal behavior. What comes out of the philosophical and psychological work on personhood are views about inalienability, uniqueness, freedom, and self-possession. The philosophical and psychological discourses that gave birth to the modern self and views of the person as endowed with the aforementioned traits also impact how we view persons, especially in political and legal contexts. Psychology and philosophical personhood have established a value to personhood in terms of identity and self-possession or ownership that is worth fighting for and protecting, that Foucault says we want to take care of. In fact, these concepts of personhood have given impetus in political struggles that sought to dominate individuals—struggles that resulted in more participatory, inclusive, and just forms of political rule, for example, the civil rights movements in various countries around the world.

Esposito reads Foucaultian governmentality with an emphasis on political life controling biological life through various biopolitical apparatuses, but one must not forget that, for Foucault, governmentality also implies a mentality of government, a governmentalization of the interior, moral self. Given that power is in a constant relation with resistance, one could argue that the governmentalization of the inner life, the moral life of the person, must also constitutionally contain sites of resistance. The building and care of the self that Foucault argues for at the end of his life admits not only powers that can control but also sites of resistance. If we view the person as a moral self, as Foucault does, we can find in this view of the moral self, always endowed with some form of moral personhood, a site of resistance to a dominant biopolitical paradigm of control, in our case neoliberal control. Foucault reminds us that the first and final point of resistance to governmentality lies in relationship we have with ourselves.[12]

At this point, we have to ask: How can the discourses of moral personhood, care of the self, and personal identity resist the political challenges posed by Esposito? What are the implications of the aforementioned discourses for Esposito's challenge to personhood? Moral personhood, care of the self, and personal identity frame the human individual in a system of values, which have practical implications for how we are to live and treat one another in society. Values such as dignity and respect for as well as the inviolability and inalienability of the person, values which are the consequences of moral personhood, personal identity, and care for the self which can motivate action aimed at preserving a perceived good, namely, the person, while giving second thought to actions and decisions that might be noxious or deleterious to the well-being of the valued person. Rights, which implicitly recognize the abovementioned values, are more than legal prescriptions. They implicitly contain and express a value system we hold dear and care about, a value system that requires a robust notion of moral personhood. When we care about these values, we will act according to them. Education is predicated on a value system, as is civic life. Aristotle remarked that friendship is more important than justice in the life of the polis. Our civic relationships with one another require more than the law or politics; they also require a deep ethical concern for the well-being of individuals and our collective life. With discourses of care and moral personhood we can establish habits and expectations, as well as set concrete boundaries regarding responsibility and accountability in society. When laws and political decisions are made that conflict with the discourses of moral personhood and care of the self, the latter discourses can rise and challenge what can be perceived to be unjust or even evil legal and political decisions, which, unfortunately, history knows all too well. To claim an insufficiency about the concept of the person is to give up powerful moral and care-full discourses that have developed over two and half thousand years of philosophy. These discourses have and will continue to motivate the desire to create a more just society, and can certainly bend to cover the failings and excesses of laws and politics. The civil rights movements over the last forty years give us plenty of examples in which moral and political discourses come into conflict and struggle, and in which the moral discourse can change and correct failings of political and legal discourses, especially on questions of race and gender inequalities.

Notes

1. "October 18, 1927, the Minister of Justice submitted a report to the Governor General of Canada regarding a petition submitted by Henrietta Muir

Edwards, Nellie McClung, Louise McKinney, Emily Murphy and Irene Parlby." The Famous 5's petition requested the Governor General to direct the Supreme Court of Canada to consider whether women were eligible to become Senators under the British North America Act, the Act of British Parliament which governed the country at this time. The Minister's report to the Governor General stated that while the government was of the view that only men were eligible to become Senators, it would nevertheless be "an Act of justice to the women of Canada to obtain the opinion of the Supreme Court of Canada upon the point." The Minister put forward the following question for the Court's consideration: Does the word 'Persons' in section 24 of the British North America Act 1867, include female persons?" From, "The Persons' Case (1929)": http://people.ucalgary.ca/~gpopconf/BASIC/person.html, accessed May 21, 2015.

2. Property and land for enfranchisement to vote. "After 1867, the colonial form of enfranchisement policy was continued by federal legislation in 1868 and then modified in 1869, so that enfranchisement and a life estate in an allotment of reserve lands could be granted to any Indian male "who from the degree of civilization to which he has attained, and the character for integrity and sobriety which he bears, appears to be a safe and suitable person for becoming a proprietor of land." http://publications.gc.ca/Collection-R/LoPBdP/BP/bp175-e.htm.

3. In various legal traditions, the notion of personhood extends to cover corporations. I do not wish here to focus on this particular aspect of personhood. Also, the person, given the context in which the concept is used, may or may not be a citizen of a state. For example, in the section just read, there is a distinction between a citizen and a person who has a status of a permanent resident of Canada. The legal force of terms like *person* and *citizen* are generally rooted either in some form of assent to a contract, this is what classical contract theorists maintain, or in more contemporary social ontological terms, thinkers like John Searle tell us these terms have force because they are performatives or as Raimo Tuomela, Margaret Gilbert, and Philip Pettit tell us, there is a rational mutual consent to the definition and use of the terms.

In my definition of the term *person*, I wish to exclude from discussion natural law or substantive accounts of the person that one finds in thinkers like Emmanuel Mounier and Jacques Maritain, which argue that there is an ontological substrate that holds all of our qualities together to form a person. I also wish to exclude from consideration metaphysical or transcendental accounts of the person that one finds in classic phenomenology in thinkers like Edmund Husserl, Edith Stein, and Max Scheler. Finally, I do not wish to follow Peter Singer's suggestion (*Practical Ethics*), which extends personhood to all forms of sentient beings and not just human beings. I exclude these notions of personhood because my focus is on how personhood is received in biopolitical debates today. This being said,

I do not deny that the various accounts of personhood just mentioned are of great political importance.

4. "The Confession of the Flesh" (1977) interview. In *Power/Knowledge Selected Interviews and Other Writings* (ed. Colin Gordon), 1980: pp. 194–228. This interview was conducted by a roundtable of historians.)

5. For example, the *Canadian Charter of Rights and Freedoms* makes the distinction between everyone, person, and citizen. See section 6 of the Charter on mobility rights, as follows:

Mobility of citizens

6. (1) Every citizen of Canada has the right to enter, remain in and leave Canada." **Also, "Rights to move and gain livelihood** (2) Every citizen of Canada and ***every person*** who has the status of a permanent resident of Canada has the right

(*a*) to move to and take up residence in any province; and

(*b*) to pursue the gaining of a livelihood in any province."

http://laws-lois.justice.gc.ca/eng/const/page-15.html, accessed May 21, 2015.

6. Roberto Esposito, "Biological Life and Political life," in *Contemporary Italian Political Philosophy*, ed. Antonio Calcagno (Albany, NY: State University of New York Press, 2015), 13–14.

7. In chapter IV of his work *Immunitas*, Esposito makes explicit the biopolitical relation between life and politics. He notes, "Politics enters fully into the immune paradigm the moment life becomes the immediate content of its action. When this occurs, all formal mediation disappears: the object of politics is no longer a 'life form,' its own specific way of being [read person], but rather life itself—all life and only life, in its mere biological reality. Whether an individual life or the life of the species is involved, life itself is what politics is called upon to make safe, precisely by immunizing it from dangers of extinction threatening it" (*IM*, 112). This citation is revealing because it makes clear certain assumptions or premises foundational for Esposito's argument. First, politics no longer only has individual personal beings as its focus: we have arrived at a point in our technologically advanced society where we have the possibility not only of destroying individual groups, systematically and efficiently so, but life itself on a global scale, be it through biological and nuclear weapons, our limited response to environmental dangers, and simply through the waste that is produced in order to keep billions of people, animals, and plants alive. We can no longer afford to think politics only at the level of individual states and communities: politics is globally more encompassing. Second, greater measures, more risky ones, are needed in order to protect life, understood in a biological sense, at this more complex and comprehensive level.

8. "This variance is an integral part of the category of person since its earliest beginnings. It has been well established that the Greek etymology of the

term refers to the theatrical mask placed on the actor's face, but precisely for this reason the *persona* was never identical to the face. The word later referred to the type of character depicted in the play, but this, too, was never the same as the actor who interpreted it from one occasion to the next. The law seems to have reproduced this element of duality or duplicity at the heart of humanity. *Persona* was not the individual as such, but only its legal status, which varied on the basis of its power relationships with others. Not surprisingly, when ancient Romans referred to their role in life, they used the expression *persona habere* (literally, 'to have a person'). *Persona* was not what one is, but what one has, like a faculty that, precisely for this reason, you could also lose. That is why, unlike what is commonly assumed, the paradigm of person produced not a union but a separation. It separated not only some from others on the basis of particular social roles, but also the individual from its biological entity. Being something other than the mask that it wore, the individual was always exposed to a possible depersonalization, defined as *capitis dimunutio*, which could go as far as the complete loss of personal identity. The category person, we might say is what made one part of the human race subject to others" (*PT*, 29–30).

9. Esposito maintains, "The disintegration process of the thing, which went hand in hand with its metaphysical treatment, now seemed inevitable. Its transposition into 'being' anticipated its formulation as 'object,' central to Heidegger's essay on the 'Age of the World Picture.' While in the Middle Ages, the thing was understood as *ens creatum*, the fruit of the creative action of God, it was later interpreted as what is represented or produced by human beings. However, by entering into the *dispositif* of representation of production, the thing—now transformed into object—was dependent on the subject, thus losing all autonomy. As Heidegger notes, this transition appears to be fully accomplished with Descartes. "What is, in its entirety, is now taken in such a way that it first is in being and only is in being, to the extent that it is set up by man, who represents and sets forth." We must not lose sight of the nexus that unifies subjectivism and objectivism while at the same time dividing them: without a subject who represents, there is no represented thing, and vice versa. The ties that the separation between persons and things seeks to sever are once again pulled tightly close together. Persons and things face each other in a relationship of mutual interchangeability: to be a subject, modern man must take the object dependent on his own production; but similarly, the object cannot exist outside of the ideational power of the subject. Kant's separation of the thing into phenomenon and noumenon—between the thing as it appears to us and the "thing in itself"—takes this splitting to its extreme conclusion. Never is the implication between the separation of the person and the disintegration of the thing so clearly visible as it is in this case. Each can be divided from the other only starting from what separates it from itself, inverting it into its opposite. Thus, while the person is always vulnerable to becoming a thing, the thing always remains subject to the domination of the person." (*PT*, 63–64).

10. Citation continues: "Thus criminals, who for other reasons are an essential element in concentration-camp society, are ordinarily sent to a camp only on completion of their prison sentence. Under all circumstances totalitarian domination sees to it that the categories gathered in the camps—Jews, carriers of disease, representatives of dying classes—have already lost their capacity for both normal or criminal action." Hannah Arendt, *The Origins of Totalitarianism* (New York: Houghton, Mifflin, Harcourt, 2001), 145.

11. Ibid., 150.

12. See Michel Foucault, *The Hermeneutics of the Subject: Lectures at the Collège de France, 1981–82* (New York: Palgrave Macmillan, 2005), 252. My thanks to Inna Viriasova for bringing this point to my attention.

7

Person and *Munus* in the Thought of Roberto Esposito

Jonathan Short

Introduction

In the context of Roberto Esposito's larger body of work, the critique of human rights featured in *Third Person* is not a call to improve liberal institutions of immunization, but rather provides evidence for their fundamental limitations. While Esposito does not suggest these limitations mean that mechanisms of immunization can be simply abandoned, he does suggest the latter's deficiencies can be tempered and opposed by developing the concept of the impersonal and the common.

In what follows here, I will attempt to show how Esposito's work on the person, the impersonal, and the common, can be mobilized to challenge an ascendant neoliberalism. I will proceed by showing, first, how Esposito unfolds a genealogy of the concept of the person and the way he believes this *dispositif* serves to undermine the tradition of human rights to which it is allegedly in service. I will then go on to suggest how we can extrapolate from Esposito's critique of the liberal concept of the person a critical assessment of neoliberalism. Here, reading Esposito alongside Foucault's work on liberal biopolitics, and following Timothy Campbell's work putting these together, I suggest that Esposito's critique of the person provides a very apt diagnosis of the way the *dispositif* of the person has become a major device of (pathological) immunity in the neoliberal era in which we live. In the second section, I seek to develop

Esposito's idea of the impersonal as a right in common in an affirmative political direction by connecting these concepts to his earlier work on the *munus* and community. In doing so, I will seek to bring Esposito's concept of an obligation held in common into dialogue with Rancière's notion of dissensus. I argue that dissensus can be grounded in the *munus* itself as a movement of expropriation counter to the immunization of some from others. Finally, I suggest some ways that this reading of *munus* and dissensus together can provide the basis for thinking the common against the neo-liberal apparatus of immunity.

Part of my argument here counters a strand of critical response to Esposito's recent work on the impersonal as an *anti-political* concept. Despite Esposito's claim that the impolitical is not anti-political, that is, an avoidance of politics, it is often thought that the impersonal or the idea of affirmative biopolitics one finds in his more recent work is an abandonment of politics or is useless for politics. I want to suggest here that not only is this not the case, but also that some of the issues with this concept are due to Esposito's own reluctance to clarify the diverse conceptual genealogy through which these concepts emerge in his work.

The Person

In his recent book *Terza Persona* of 2007, Esposito introduces the idea of a "*dispositif* of the Person" to extend his previous discussions of the concept of immunization first explored in relation to community (*CM*) and biopolitics (*IM* and *BP*). According to Esposito, the *dispositif* of the person has the effect of placing human beings into different categories of worthiness for the protections afforded by human rights. This feature is what, to him, prevents the doctrine of universal human rights from actually applying to all human beings. Even though human rights are designed to protect (immunize) all human beings from violence, those rights are predicated in turn on the concept of the person. This category is established through an internal distinction between *homo* and *persona*. It functions as though there were two interlocking strategies of immunization at work simultaneously. Because of its dual mechanism—functioning in a similar way to a negation of negation—violence against human beings is both forbidden and permitted. The immunizing function of universal rights against violence is itself subject to a counterimmunization because, while all human *persons* are certainly entitled to the protections afforded by right, not all human beings count as persons. According to Esposito, the universal pretensions of the institution of human rights are thereby compromised in advance by the need to specify which living beings qualify for personhood.

That the category of the person is split in the manner just described is neither a simple historical accident nor can it be a mere juridical oversight. As Esposito argues in the introduction to *Third Person*, the *dispositif* of the person emerges in ancient Roman legal codes already imbued with the strategic function of creating different classes of people based on their distribution between the poles of person and (animated, natural) thing. The status of personhood, Esposito notes, is "only an interlude, a sort of unnatural pause on the servile horizon that included within its larger compass all human beings—with the exception of adult male Roman citizens" (*TP*, 10). As he argues in another essay recapitulating the main analysis of *Third Person*, the conception of personhood in the context of the Roman *Imperium* functioned specifically as an apparatus for appropriating the lives of others. Citing Simone Weil, Esposito notes that according to her, "the bridge between Roman law and violence is constituted by property: owning things and men transformed into things through the institution of slavery constitutes not only the context of the juridical order, but its form."[1] This is in fact why, rather than producing completely separate categories, the Roman legal paradigm casts *homo* and *persona* into a set of mobile designations allowing for a whole series of intermediate gradations between the two categories—for instance, the slave who can enter into contracts on behalf of the master as much as the child who is subject to the unlimited authority of the *pater*. The upshot of this mobility, however, is to consolidate the domination of the few legal persons over the many servile human beings by judging the latter to be types of instruments or forms of property entirely dependent on the former. Thus, since the designation of person perpetually plays opposite the natural human being, the category of rights that attend the person are also those of appropriating and securing property, establishing a relation of domination between the person and what is appropriated out of the "things" of nature, living and non-living.

For Esposito, the internal tie between the person and the institution of property continues in a changed but still recognizable form right through into modern liberal society. If medieval Christianity reacted to the brutality of the *Imperium* by declaring all human beings to be persons, this does not effectively stop the division internal to the category from operating. Remaining indebted to the metaphysics inherited from the ancient world in which the soul is closer to the divine than is the body, Christianity, perhaps despite its best intentions, reproduces the duality that is "put together in such a way that one of its elements is subordinated to another, separating it from God."[2] It is this element of subordination—or as Esposito will also argue following Foucault, subjection—of corporeality to the soul, that liberalism will in its turn take up in a modified form. In this context Esposito references the arguments of Locke and Mill that "the body is owned by the person who dwells inside it" (*TP*, 12). In liberal thought, the rational agent calculating its advantage is regarded as distinct from that same agent's natural-corporeal qualities, which the agent now appropriates and disposes of as it sees fit. The economic subject, as much as the juridical person of liberalism, logically presupposes a hierarchical split between mind and body, wherein the latter becomes "an appropriated thing"; hence "the person is specifically defined by the distance that separates it from the body" (*TP*, 13). Uniting the ancient difference between *homo* and *persona* in one personal/impersonal body, in effect becoming its own property, the self-owning person of liberalism becomes the conduit for the appropriation of internal and external nature.

It is above all on the modern biopolitical horizon outlined by Foucault[3] that the liberal paradigm of self-ownership becomes key to understanding the modern present. But instead of following Foucault's suggestion that we find an irreducible difference between the ancient (sovereign) power that takes life and lets live and the biopolitics of the population that expressly mandates life while only sometimes disallowing it, Esposito maintains these two strategies of power are closer than they might at first appear. Thus Esposito argues that in the modern liberal context, we find affirmed once again the "ancient Roman separation between *persona* and *homo*" which based on judgments of the value of the living "marks the final difference between what must live and what can legitimately be cast to death" (*TP*, 13).

While this is not to deny that there is a gap between the Roman *dispositif* and the one of contemporary liberal-governmental biopower, it is to argue that this discontinuity is linked in a way that calls into question the assertion of a substantive break between modernity and what came before it. According to Esposito, it is in fact historical discontinuity itself that provides the matrix for a sort of return of the archaic in the midst of contemporary circumstances that appear very different from it. As he argues, "it is the breaking of chronological continuity [. . .] that opens up, in the flux of time, those empty spaces, those fractures, and those crevasses in which the archaic can once again re-emerge."[4] Such reemergence, to be sure, is never the simple repetition of the archaic form per se, but appears rather "as a specter or phantasm," manifesting in history much the way Freud thought that psychic phenomena most strongly rejected were the very ones most susceptible to repetition.[5]

From this perspective, developing the proximity Esposito sees between ancient and modern forms of biopolitics, Timothy Campbell has offered an intriguing set of suggestions for how to use Esposito's discussion of the person to analyze contemporary neoliberalism.[6] According to Campbell, the notion of self-ownership Esposito identifies as the hallmark of the liberal person is intensified and expanded by neoliberalism. This means primarily that the distance between person and human being (as natural thing) is increased to the point where the relations between them threaten to make each side indistinguishable from its opposite. In contemporary neo-liberal biopolitics, it is no longer simply the trait of self-ownership that supplies the distinction between person and non-person, but rather, it is "a person's capacity to increase her biopower [that] will become the primary means for determining how fully she is a person."[7] As we have seen, prior forms of liberalism instrumentalize the body so that both it and its capacities could be turned into commodities for sale on the capitalist market.[8] In this sense, liberalism creates the conditions for a biopolitical regime in which developing and maximizing individual capacities is equated with the maximization of freedom, with the consequence that the latter is conceived in entirely individualistic terms.

Neoliberalism, for its part, will intensify this dynamic, not just by demanding that individuals increase their productive biological capacities, but by setting up a specific social regime that enshrines wholesale competition and self-management as the normative model for all spheres of life. In response to this competition, individuals have little choice but to maximize their biopower on pain of falling behind in the competitive struggle. To fall behind is not merely to risk falling into poverty, but in doing so the subject's status as a person is also degraded and put at risk. The subject's status as a person is therefore dependent on a set of competitive performances, failure at which strongly suggests a lack of those qualities—initiative, the willingness to take risks, self-discipline and a resilient work ethic—that make one a person.[9] Indeed, to lack personhood in neoliberalism, unlike in earlier versions of liberalism (for example, the welfare or social democratic liberalism of the postwar period), is to bring to bear on the faulty person a regime of disciplinary control over that individual's life, or even worse, to consign them to one of the growing "zones of abandonment" to which the losers of the competitive game are increasingly consigned, and where they are subject to an apparatus of thanatopolitics.[10] As Campbell's discussion makes clear, Esposito's work on the person in its modern and archaic varieties joins Foucault's earlier remarks on the modern regime of biopolitics with the latter's lectures on neoliberal governmentality, not coincidentally titled *The Birth of Biopolitics.*[11]

As Foucault argues in his *Birth of Biopolitics* lectures, much of the intensification of early liberalism has taken place through the notion of the "enterprise self," which, following Esposito's analysis, should be considered a new twist in the *dispositif* of the liberal person in neoliberal societies. According to Foucault, in a crucial section of his lectures, the self-owning person of classical liberalism becomes in neoliberal economic theory a self-relation of human capital. We might define this self-relation precisely as a type of self-objectification in which the subject turns his or her capacities into a sort of capital to be rationally managed, maintained, and enhanced. Such a view of the (economic) subject presupposes the notion of the person Esposito claims constitutes the liberal subject, in other words, a subject not only relating to itself in the mode of self-ownership, but furthermore, calculating on the basis of this ownership how to maximize a return (an income), from its own utilization. Since this ownership exists for the express purpose of calculative maximization, everything that belongs to this person can be included as its capital. Hence, as Foucault summarizes the argument of those neoliberals who he calls here "neo-economists," theirs "is not a conception of labour power; it is a conception of capital-ability which, according to diverse variables, receives a certain income that is a wage, an income-wage, so that *the worker himself appears as a sort of enterprise.*"[12] Neatly undercutting the Marxist distinction between capitalists who employ labor power and workers who sell it, the neo-economists subsume the latter into the calculative matrix previously reserved for the former. We might also say, by drawing on the liberal *dispositif* of the person, that the neoliberal economists draw all social actors into a kind of universalizing economic reason of calculative maximization. In other words, there is a direct relationship between the liberal person regarded as self-owner and the universal economic agent conceived as an "enterprise." Thus for neoliberal economics a global grid of intelligibility takes shape where anything an individual does can be analyzed in terms of benefits and costs relative to improvement or deterioration of the entrepreneur-subject and its capital.[13] Under this *dispositif* is constituted what Foucault defines as "*homo oeconomicus* as entrepreneur of himself, being for himself his own capital, being for himself his own producer."[14]

What must be added to this brief discussion of neoliberal economic thought is the historical-political project transforming neoliberalism from a body of economic thought into a form of life, which also transforms the way subjects conceive of themselves. In an ascendant neoliberal society, as Louis McNay has observed, Foucault's own notion of "care of the self," initially a form of resistance and refusal of forms of biopolitical governance, threatens to become indistinguishable from the self-entrepreneurship characterizing the neoliberal person.[15] Indeed, as Dardot and Laval have recently maintained, "Life itself, in all its aspects, becomes the object of apparatuses of performance and pleasure" such that the subject/person in its entirety becomes the field of a potential submission to power.[16] This point recalls Campbell's idea that what is specific to the neoliberal *dispositif* of the person is that the more fully the subject maximizes their biopolitical capital, the more they are considered persons. The person thus becomes an objectified and internalized grid of idealized performance to which actual subjects must continually measure up on pain of failure and drift toward the pole of thing. The individual, now entirely responsible for his or her fate, risks joining the increasing numbers of the population who, having little or no chance of economic success, take on the social position of disposable things. In this situation, neoliberalism can be said to reproduce in a new form the shift between person and animate thing whose contours Esposito traces beneath the liberal utopia of universal human rights.

Following the analysis of Dardot and Laval, we might even push this point a little further. As they maintain, unlike previous regimes of liberalism, the ideal of the neoliberal person is of "an *ultra*-subjectivation, whose goal is not a final, stable condition of 'self-possession,' but a beyond the self that is always receding, and which is constitutionally aligned in its very regime with the logic of the enterprise," and beyond it, to the inscrutable and uncontrollable vicissitudes of the capitalist market.[17] In point of fact, Foucault already registered that the subject who responds rationally and systematically to an external environment not under their control becomes "eminently governable," rather than being the free agent posited by liberal economic theory.[18] What Foucault perhaps did not see as lucidly as Deleuze did,[19] however, was that biopolitical power was increasingly to take the form of a network of control uniting in a single apparatus the disciplining of the body and the management of populations typical of the networked, digital capitalism that has become synonymous with the neo-liberal era. To re-introduce Esposito's perspective on the person, the neoliberal person threatens to make the difference between person and thing indeterminate. If the person is the one who responds in a systematic way to the forces of the market, while at the same time can never hope to control these forces on the individual scale of the enterprise self and its human capital, one might justly assume that such a person is but a calculating, animated thing; here, the latter's freedom is reduced to the calculation required to stave off an even greater level of servility and indignity. But at no time does the neoliberal subject escape a kind of dependence on the outside that dictates its responses, and so, is not that different from the merely human life its calculations were to allow it to escape.

The problem, for Esposito, is that the person is a device of immunity. As such, it shares in the common defect Esposito ascribes to that paradigm in general, which is that immunity—at least past a certain point—becomes pathologically self-destroying. As he writes in *Immunitas*, "the immune mechanism functions precisely through the use of what it opposes. It reproduces in a controlled form exactly what it is meant to protect us from" (*IM*, 8). But for Esposito, this mechanism becomes pathological when it becomes excessive and uncontrolled, as in certain forms of auto-immune disorders in which the body is not able to distinguish between itself and the foreign body that it seeks to neutralize, destroying itself in the process. In this context, the rights-bearing person is immune to the dangers that affect those who do not have such protective rights, those who are non-persons or living things; and yet, the status of the person in neoliberalism entails a punishing quest for total maximization that not only renders increasing numbers ineligible, but also makes those who qualify only temporarily immune from the status of living thing, to which, not unlike in ancient Rome, they are always in danger of returning. It is true, of course, that great wealth can provide a powerful buffer against such contingency, and yet there are no guarantees that such wealth will not be lost amidst the increasing volatility and competition of global capitalism.[20]

The Impersonal as a Right in Common

What is at stake for Esposito in his discussion of the impersonal is a form of right that is not predicated on the status of the person. His attempt to introduce a concept of the impersonal in the concluding section of *Third Person* includes statements that suggest the possibility that right can be thought of in some other way than as "personal" right. It is this suggestion I will now try to develop by linking the concept of the impersonal to Esposito's earlier work on community.

Paraphrasing Simone Weil, Esposito writes that "What is sacred in humans is not their *persona*; it is that which is not covered by their mask. Only this has the chance of reforging the relationship between humanity and rights that was interrupted by the immunitary machine of the person" (*TP*, 16). Developing this idea, Esposito suggests that what we require is "something as seemingly contradictory as 'a common right' or a 'right in common' " (*TP*, 16). Subsequently investigating a series of figures of the impersonal as prospective paths to an "affirmative biopolitics," Esposito argues that these compose a thought of a "third" positioned on the margins of the person but which can never be assimilated to the person. The impersonal is, he writes, "a point, or layer, which prevents the natural transition from the splitting of the individual—what we call self-consciousness or self-affirmation—to the collective doubling, to social recognition" (*TP*, 102).[21] Seemingly impossible, the impersonal connects the unique and singular aspect of each human being with what is common to all, corresponding to "the rights of the entire human community" (*TP*, 103).

What this suggests is that Esposito is not completely hostile to the concept of right per se, nor does he view the concept of right as inherently serving to immunize each from all or some from others. This is in a way not too surprising because Esposito's earlier work on the origins of the concept of the community in the ancient *munus* seems to authorize something like this paradoxical right that can only be thought as an obligation or an expropriation common to all, that is, a common right that is held equally by all rather than a right that one holds against someone else or against the collective. Accordingly, for Esposito in *Communitas*, "What predominates in the *munus* is [. . .] reciprocity or 'mutuality' (*munus-mutuus*) of giving that assigns the one to the other in an obligation [*impegno*]" (*CM*, 5). Here, the *munus* is what obligates one to give, it is what one owes the community because one has received from it what is principally *not* one's own, that is, the gift of life itself, which is impersonal and cannot be said to be the personal property of any living being. This is despite Esposito's claim in *Third Person* that what is common is also unique to each one, since the force of his argument is that what is unique is not a property of our *person* but rather consists in the singularly *impersonal* features of life that we did not or could not have chosen, and which is not ultimately under our control or up to us (*TP*, 104). Thus, the impersonality of the life of the living beings that we are is what cannot be appropriated or assimilated but on which we—even as persons—do not cease to depend. This means that one's living is never one's property, but rather constitutes a radical impropriety, a lack of property that is the breach or gap through which the community lives in us and so disposes and obliges us toward it.

Because each is in this position of impropriety or dependency with respect to the community, a de facto obligation is imposed on each to which each is bound. Thus Esposito writes, "The subjects of community are united by an 'obligation,' in the sense that we say 'I owe you something,' but not 'you owe me something' " (*CM*, 6). The important insight here is that the paradoxical "common right" at issue is not something owed the individual by the community (the traditional notion of right), but rather the reverse, that the community is owed by the individual. Yet this is not a doctrine of the tyranny of the collective over the individual because each subject of the community shares this owing in equal measure. Although the debt is experienced as non-reciprocal, as something one owes rather than what is owed someone, it is nevertheless an entirely mutual indebtedness, and so no one is immune from the obligation it imposes. This is why Esposito can argue that the community "isn't the subject's expansion or multiplication but its exposure to what interrupts the closing and turns it inside out: a dizziness, a syncope, a spasm in the continuity of the subject" that is the anonymous being of community subsisting through the individual (*CM*, 7). Such a community without immunity is also a community without divisions and differences; an undifferentiated mass, it cannot tyrannize over its members except to the degree the tyrannized and tyrannizing are the same. This community does not yet contain the differentiation necessary for some to dominate others on the basis of differences and the ranking of these differences according to some value scheme.

It is impossible not to notice that Esposito derives the poison and its antidote from the same singular historical source, that is, ancient Roman society. As he derives the *munus* from the same society that invented the person for purposes of immunity, it is as though Esposito were attempting to uncover the primitive foundations of a pre-political collective beneath the layer on which the political apparatus of the person was subsequently founded.[22] This seems to reflect at least a logical, if not temporal, priority, because the *munus* logically precedes the apparatus of the person that introduces an immunizing function and exemption from a prior mutual obligation. The *communitas* would then be an image of a social body at the zero degree of politics, where all are mutually obligated to give, each being in debt to the rest for their very lives, and yet, precisely for this reason, where no one is in a position to withhold anything as immune to expropriation. If this is indeed an archaic image of a "primitive communism" seemingly prior to politics, it is one founded on complete indifferentiation, on the impersonal in the human being that all human beings share in common. If such a "society" never actually existed, nor could ever persist in a stable social form, it nevertheless seems to trace the outline of an original social condition subtending the historical politics of immunity. The latter would necessarily be predicated on the hierarchical distribution of differences according to some value scheme, and so the rule of some over others, the distribution of individuals and populations into persons and non-persons with which we are familiar.

However, what I want to argue is that the image of the *munus* as somehow pre- or a-political is mistaken; it is not a useless concept for politics. The *munus* is actually the ground of politics, its ultimate stake that cannot be reduced to one of the values supportive of a schema of value. It is this irreducibility that comprises the political effect of the *munus*. Whatever the historical status of the *munus*, by recalling it Esposito is not engaged in an exercise of nostalgia or an attempt to rediscover a lost domain of pre-political innocence. Quite the opposite, as he explicitly claims, the *munus* should be understood to give to collective obligation a new political sense.[23] It is the inevitable persistence of the *munus* in the contemporary political situation, composed of highly immunized persons, that serves as a basis for the formation of political subjects whose role is precisely to contest the distribution of immunity in the apparatus of the person.

At this point we can turn to another attempt to contest the political distribution of persons and rights, as is found in the work of Jacques Rancière. In particular, Rancière's discussion of human rights in his important essay "Who is the Subject of the Rights of Man?" provides a point of comparison between his own concept of dissensus and Esposito's *munus*. Rancière defines dissensus as "part of the configuration of the given, which does not only consist in a situation of inequality, but also contains an inscription that gives equality a form of visibility."[24] My argument is that it is possible to see the *munus* in terms that are not only very similar, but which reveal the political stakes it presents.

In his essay Rancière shows political dissensus in a way that is telling for bringing out the *munus* as a compositional moment in political conflict. Rancière cites the claim of Olympe de Gouges during the French Revolution that "if women were entitled to go to the scaffold, then they were also entitled to go to the assembly."[25] Rancière interprets this statement to mean that the women in question "acted as subjects that did not have the rights that they had and that had the rights that they had not."[26] Rancière is absolutely correct to argue that this claim is not merely (although it also is) an assertion of women's belonging *de facto* to an order from which they were excluded *de jure*. De Gouges is not merely claiming entitlement to the "rights of Man," which is to say full personhood, on the basis of what Rancière describes as "a conflict of interests, opinions or values."[27] Rather, and more fundamentally, de Gouge's statement introduces dissensus, "a division inserted in 'common sense' " in the existing distribution of political subjects and rights.[28] This division contests the way the existing order of qualification and disqualification operates by asserting that that order is blind to the very equality it presupposes.

But which equality is that? In his essay Rancière draws attention to de Gouges's reasoning that if any subject could be obliged to lose their "bare life" for the sake of the community, then all such subjects are already obligated at the most fundamental level, and are thus already included in the order from which they are nonetheless excluded. In other words, if women were "as equal 'as men' under the guillotine, then they had the right to the whole of equality."[29] The precise point of the dissensus thus concerns what it is that should determine membership and inclusion. De Gouges does not contest the obligation imposed by the community. She does not dispute the idea that her life is not merely her own but also can be demanded by the community. In fact, she explicitly invokes this right of the community, the obligation it imposes, in two related ways. First, she argues that the obligation itself is the fundamental determinate of inclusion because it is an obligation imposed on all; the common or mutually held obligation is the one that should serve as the criteria of membership. Second, she asserts that fact of common obligation in order to depose or devalue the competing value scheme under which she is (as are all other women) excluded from the assembly. The assertion of right is not just a bid for membership in the extant value scheme as it stands, but also a contestation—a dissensus—over the criterion determining its applicability. And this dissensus takes the form of an attack on a narrower definition of inclusion on the basis of the equality—that of common obligation—that underwrites the entire situation but is denied its visible place.

What this discussion demonstrates, in my view, is that the *munus* as Esposito understands it, is not apolitical, but serves as the political "ground zero" of claims to equality and membership. In the situation described by Rancière, it is this common obligation that is invoked by de Gouges as the presupposition of equality and the dissensus that contests her exclusion from it by displaying or making that equality apparent. The *munus* at stake in this dissensus is not an abstract or even "vitalist" force, but rather shows up through the claim to an equality of obligation that can be made apparent whenever a situation of inequality is asserted on some other basis than that of the common itself. While nothing guarantees that it will be made apparent in an actual political and historical situation, that dissensus on the basis of the *munus* will actually occur, it remains a kind of potential that can appear politically into the social order from the situation of commonality and of human beings' dependency on the common.

It might be said by way of objection that the *munus* lacks historical specificity, so that appealing to the obligation to give one's life to the community is relevant only in some (extreme) cases, and so it would otherwise be an abstraction. But this is missing the point. The claim of the *munus* is the obligation to give based on dependency on the common. The obligation to give can and will take different historical forms: in the case of revolutionary France, it is the obligation to die for the community seeking to rid itself of a whole layer of immunity in the form of nobles and clergy; in our present circumstances it might rather appear as the imperative to contest the power of corporations and the inequality that excludes so many from the very community of which they are part. But whatever the case, the *munus* takes the form of an imperative toward equality against immunity. We can even say it is likely to appear wherever immunity has reached the point of pathological insulation, attacking the very community on which it depends.

At this point it should be noted that the political potential of the *munus* is often overshadowed by Esposito's more recent work on the impersonal as taken up in texts such as *Bios* and *Third Person*. In these texts, we find the impersonal juxtaposed to the personal, and from this perspective, it looks like an appeal to an anonymous vital force of life within each person that in a rather unspecified way subverts the apparatus of the person. Now, as a series of critics have contended (at one point myself included), if this is how the impersonal is read, it is hard to know what to do with it politically—it cannot be a *subject* of politics since it is precisely what the subject depends on without ever being able to be appropriated. It also seems that the commonly anonymous fact of life, the fact that all living beings share "life" as their common condition, is not enough to constitute a community or generate obligation as I have argued the *munus* does. So if we read the impersonal as something like the common fact of a (separated) force of life inherent in each individual, the criticisms of an affirmative biopolitics as a non-political concept seem justified.[30]

Thus, from the perspective I am trying to present here, it seems that Esposito has articulated the impersonal in an ambiguous fashion. One strand, the one through which he developed the concept of the *munus*, comes via Jean-Luc Nancy's attempt to rethink Heideggerian *Dasein* through the concept of *partage*, the sharing (out) of being. Esposito gives to this concept of being in common a sense of the being of the common through the *munus*, pushing it in a political direction that he argues is absent from the work of Nancy. This strand is clearly visible in *Communitas* and elsewhere.[31] However at the same time, Esposito also draws quite consciously on Deleuze and Bergson in order to develop a concept of the impersonal that appears much more closely aligned to a vitalist tradition in contemporary thought. In many of these same texts just cited, Esposito already appeals to this tradition to ground an affirmative biopolitics in the concept of the impersonal. Esposito, at least to my knowledge, has never attempted to reconcile or sort out the implications of drawing on these two strands, and this has the unfortunate effect of obscuring the political import of his critique of the person. What I have tried to do in this section, then, has been to show how the impersonal can be read through the *munus* in a political direction.

For Esposito, the point of developing the concept of the *munus* as a way to rethink the common is precisely to allow for a renewed thinking of the common good. As I have argued elsewhere, this commitment situates Esposito in a republican political tradition seeking to think freedom and obligation together through life lived in common.[32] Reimagining the common as common good is about more than reasserting the dignity of public space, although Esposito believes the latter should not be abandoned to the extant dynamics of neoliberal privatization. The public, as the sphere of the state and government, does not exhaust the common, since the latter can never be reduced to the sphere of the state's institution of the *res publica*. The problem is, Esposito writes, that "[t]he common is neither the public—which is dialectically opposed to the private—nor the global, to which the local corresponds. It is something largely unknown, and even refractory, to our conceptual categories, which have long been organized by the general immune *dispositif*."[33] In any case, Esposito's work on the possibilities of the common should not be seen as simply without political import. Rather, precisely by linking immunity to the private, the proper, and the person, as opposed to the common, Esposito's work seeks to formulate a new (or renewed) form of political engagement.

Conclusion

Neoliberalism is a regime of immunity in which, to use Esposito's and Foucault's terms, the *dispositif* of the person is deployed in order to turn subjects into "enterprise selves" who adapt to the demands of a highly competitive form of globally networked high-technology capitalism. By turning the *dispositif* toward a self-maximizing subject of continual productivity, it forces subjects to continually change, but to do so in a way that imprisons the dynamic of innovation in the form of competitive adaptation. Change is thus captured in a way that seeks to reproduce the same. Thus, despite neoliberalism's assertions about its own dynamism, it resembles closely Rancière's description of a community envisioned along the lines of consensus.[34] Consensus for Rancière entails a model of the community in which all are allotted a place, all are counted, and nothing is permitted to be out of place. If we have witnessed a decline in the relevance of politics under neoliberalism—until perhaps more recently—this is because, as Rancière would argue, neoliberalism represents a consensus-form in which the dissensus of the community is managed and subjects are governed so as to allow the existing order to be reproduced with minimal disruption.

For Rancière, however, community can be conceived other than through consensus, and this way is that of dissensus. In dissensus, as we have seen, one finds "*surplus* subjects that inscribe the count of the uncounted as a supplement" to the police consensus of parts.[35] For Rancière, this "part that has no part . . . acts to separate the community from its parts, places, functions and qualifications."[36] Dissensus asserts the truth that "politics is a process, not a sphere."[37] If political activity has been reemerging from its long slumber in recent times, if dissensus appears once again to be possible, we might say that neoliberalism produces an excess of non- or failed persons who not only become uncountable—for example, those who inhabit Campbell's "zones of abandonment"—but whose very uncountability returns to contest neoliberal countability (and accountability) as such.

As I have tried to sketch in this piece of writing, it is here at the point of dissensus that we can find a place for Esposito's concept of the *munus*. The return of politics is connected under neoliberal hegemony with a popular assertion of the value of equality. Appeals to equality are simultaneously appeals to the notion of responsibility and obligation, the idea that the enormous immunity of the very wealthy to the community on which they depend violates a fundamental obligation imposed by the existence of community itself. In our times, politics takes the form of seeking to make the immune persons obligated once again. And much of this demand or dissensus is articulated by those who have been rendered politically invisible, in Rancière's terms, uncountable, in asserting their (denied) equality and hence right as political subjects of community to impose a communal obligation on the immune.

In fact we might say that the *munus* of Esposito is actually the condition for the continued relevance of the paradigm of right, and that right must be conceived not as providing immunity but as intending the continuation of the common. In this sense, politics and the demand for rights on the part of those deemed less than persons or surplus living beings is the moment of reasserting the *munus* of common obligation that troubles and makes contestable the given order of spheres, persons, and rights.[38] What this suggests is that all rights are already common right by virtue of the fact that, as Rancière puts it, the "strength of those rights lies in the back-and-forth movement between the initial inscription of the right and the dissensual stage on which it is put to the test."[39] It is in this sense also, then, that we can agree with Esposito that the common is not the same as the public, even if this should not entail the notion that the public is not also crucially important.

Notes

1. Roberto Esposito, "The Dispositif of the Person," *Law, Culture and the Humanities* 8, no. 1 (2012): 29.

2. Ibid., 21.

3. Michel Foucault, *The History of Sexuality, Volume One: An Introduction*, trans. Robert Hurley (New York: Pantheon Books, 1978).

4. Esposito, "The Dispositif of the Person," 27–28.

5. Ibid.

6. Timothy Campbell, *Improper Life: Technology and Biopolitics from Heidegger to Agamben* (Minneapolis: University of Minnesota Press, 2011).

7. Ibid., 74.

8. *IM*, 13; Campbell, *Improper Life*, 74.

9. Campbell suggests that the signs of grace Weber equates with charismatic authority have actually been transposed into the fate of individuals on the capitalist market (Campbell, *Improper Life*, 69–70).

10. Ibid., 75–76.

11. Michel Foucault, *The Birth of Biopolitics: Lectures at the College de France 1978–79*, trans. Graham Burchell (New York: Palgrave McMillan, 2008).

12. Ibid., 225; added emphasis.

13. Foucault primarily has in mind here the influential work of Gary Becker, who he refers to several times in these pages.

14. Foucault, *The Birth of Biopolitics*, 226.

15. Lois McNay, "Self as Enterprise: Dilemmas of Control and Resistance in Foucault's *The Birth of Biopolitics*," *Theory, Culture & Society* 26, no. 6 (2009): 55–77.

16. Pierre Dardot and Christian Laval, *On the New Way of the World: On Neoliberal Society* (London: Verso, 2014), 283.

17. Ibid., 284.

18. Foucault, *The Birth of Biopolitics*, 270.

19. See Gilles Deleuze, "Postscript on the Societies of Control," *October* 59 (1992): 3–7.

20. One might suggest that the primary utility of such wealth is access to the political power necessary to prevent such contingency.

21. The problem with this formulation, as I take up below—and why I seek to connect the impersonal to the *munus*—is that the impersonal, if it is to mean anything, must be (potentially) socially recognized. What I seek to show below is that dissensus is precisely the attempt to make the impersonal socially recognizable, in other words, that it becomes a political stake.

22. Esposito had defined the concept of immunity in *Communitas* as a property of the subject allowing exemption from communal obligation, allowing that subject to "completely preserve his own position" (*CM*, 6).

23. Roberto Esposito, "Community, Immunity, Biopolitics," trans. Zakiya Hanafi, *Angelaki: Journal of the Theoretical Humanities* 18, no. 3 (2013): 84.

24. Jacques Rancière, "Who Is the Subject of the Rights of Man?," in *Dissensus: On Politics and Aesthetics* (London: Verso, 2010), 68.

25. Ibid.

26. Ibid., 69.

27. Ibid.

28. Ibid.

29. Ibid.

30. If the impersonal is tied to the *munus* in the way I have suggested here, this allows us to temper the criticisms Esposito's work has generated equating the impolitical with the non- or anti-political. See Bruno Bosteels, "Politics, Infrapolitics, and the Impolitical: Notes on the Thought of Roberto Esposito and Alberto Moreiras," *CR: The New Centennial Review* 10, no. 2 (2010): 205–38; Peter Goodrich, "The Theatre of Emblems: On the Optical Apparatus and the Investiture of Persons," *Law, Culture and the Humanities* 8, no. 1 (2012): 47–67; Jonathan Short, "On an Obligatory Nothing: Situating the Political in Post-metaphysical Community," *Angelaki: Journal of the Theoretical Humanities* 18, no. 3 (2013): 139–54; Matheson Russell, "The Politics of the Third Person: Esposito's *Third Person* and Rancière's *Disagreement*," *Critical Horizons* 15, no. 3 (2014): 211–30. In the case of Russell, who also compares Esposito and Rancière, his argument that the impersonal in Esposito consists of a mere "appeal to the outside" that disrupts, yet without ever articulating positively, existing social categories, holds only if the impersonal is not brought into contact with the *munus* (Russell, "The Politics of the Third Person, 221). Russell quotes an earlier paper of mine criticizing Esposito's affirmative biopolitics in this sense; my argument in this paper should thus also be read as a rejoinder to my own earlier position on the impersonal. The mode of the impersonal found in the *munus* is not outside the community, and so it canot be thought of in terms of an ethical gesture to something "beyond." Rather, the *munus*, while impersonal, provides a political dimension to claims against immunity that can be taken up by subjects in the name of a shared obligation to the gift of the common.

31. Esposito, "Community, Immunity, Biopolitics;" Roberto Esposito and Jean-Luc Nancy, "Dialogue on Philosophy to Come," *The Minnesota Review* 75 (Fall 2010): 71–88.

32. Jonathan Short and Gregory Bird, "Community, Immunity, and the Proper: An introduction to the Political Theory of Roberto Esposito," *Angelaki: Journal of the Theoretical Humanities* 18, no. 3 (2013): 1–15.

33. Esposito, "Community, Immunity, Biopolitics," 89.

34. Rancière, "Who Is the Subject of the Rights of Man?"

35. Ibid., 70.

36. Ibid.

37. Ibid.

38. Foucault says something quite similar toward the end of Volume I of the *History of Sexuality* where he invokes popular agitation against governmental forms of biopolitics to a paradoxical set of rights, such as to health or life, having no precedent in the rights tradition because these are not primarily individual rights.

39. Rancière, "Who Is the Subject of the Rights of Man?," 71.

8

From Biopolitics to Political Animism

Roberto Esposito's Things

Federico Luisetti

The thingness of the thing remains concealed, forgotten.

—Martin Heidegger[1]

Roberto Esposito is known primarily for his rethinking of philosophical categories from the perspective of the "impolitical" and his contributions to the biopolitical debate inaugurated by Michel Foucault.[2] More recently, his critique of political theology has led him to an ambitious confrontation with Western political philosophy and revitalization of Renaissance naturalism and contemporary vitalisms.[3] His recent book, *Le persone e le cose* (*Persons and Things*),[4] sums up and prolongs this trajectory, while at the same time suggesting structural convergences and alliances with ethnographic, postcolonial and science studies approaches.

In the following pages I will focus on the key arguments of *Persons and Things* and frame them in the context of contemporary debates on naturalism and the politics of life. In my opinion, Esposito's thought can be approached as an internal deconstruction of the "state of nature" devised by Western modernity and an attempt at provincializing the Euro-Atlantic legacy of political philosophy.[5] From this perspective, Esposito's texts resonate with decolonial, posthuman, and anthropocenic discourses. In all such instances, as in Esposito's anti-theologico-political critique of Western modernity, subjectivity is thought without reference to the interiority of a human subject, as a function of things and as a form of corporeal agency.[6]

Esposito's fierce polemics against the hegemony of juridico-political dualisms in modern philosophy, his critique of the persons/things divide and the Western apparatus of ontological subjectivism and idealism, his recovering of radical naturalisms, body-politics, and the *nexum* that in ancient Roman law and premodern societies links subjective functions to the life of things, separating them from the person and the human, can thus be seen as a philosophical project that reconfigures the Eurocentric state of nature. *Persons and Things* points at these convergences with anthropological thought and decolonial studies, overlapping the destitution of philosophical modernity with the recuperation of premodern paradigms still active within our actuality, as suggested by the work of Marcel Mauss and Bruno Latour (*PC*, 95–99). Since theoretical alliances presuppose also divergent philosophical declinations and geopolitics of knowledge, the following pages underline the tensions around the understanding of modernity, naturalism, and coloniality, and the challenges entailed by the inscription of European biopolitics within the state of nature of Western modernity.

Res/Persona

Although conceived from within European philosophical thought—as a close-knit dialogue in the history of ideas with Kant and Heidegger, Descartes and Spinoza, Bergson and Merleau-Ponty, Hobbes and Weil, Hegel and Foucault—*Persons and Things* quietly unsettles some of the presuppositions of Western philosophy. Its central thesis is that a long-term juridico-theological logic, codified by Roman law and Christian theology, has oriented Western episteme, separating things from persons, and leaving the body as an unthought, ambiguous, and excluded dimension of human experience. From Esposito's perspective—which is methodologically akin to Agamben's theologico-political archaeology of actuality—modernity must not be regarded as a radical break with the past but as an intensification of ancient premises, shaped by the composition of Greek philosophy, Roman law, and Christian thought (*PC*, ix).

Juridico-philosophical thought has programmatically denied the corporeal dimension, letting power and politics function silently on the bodies of its subjects, reproducing within the political body the *summa divisio* of Gaio's *Institutiones* and separating hierarchically persons from things, freedom from slavery, the head from the other organs, the soul from the body, as in Hobbes's *Leviathan*. The juridical apparatus of separation, this theoretical *dispositif* of unification by means of the subordination of an inferior part and exclusion of a repressed dimension, has faithfully reproduced a political articulation of purified polarities that has been functioning for centuries through a variety of oppositions: the two bodies of the king, sovereignty and the people, charismatic domination and legal authority, personal power and impersonal regulations.

Esposito's argument is that our biopolitical and biotechnological present is finally exhausting this conceptual apparatus, so that the task of contemporary philosophy is to accompany the "sunset" of the "horizon of modernity" (*PC*, ix) and offer new categories for interpreting a radically new reality. No longer, as Foucault has intuited, the dualistic juridico-political vocabulary is able to grasp the dynamics of contemporary societies. A biopolitical transition, a radical mutation led by "globalization and technological revolution" (*PC*, 109), a historical discontinuity that has transformed forever political governance and representation, social desires and forms of production, is forcing thought to get rid of its assumptions. Masses, people, multitudes: no name adequately describes the formless political bodies of contemporary man and women.

Esposito's insistence on the fracture between experience and thought is apparently reprising a quietist Hegelian motif: "philosophy always comes too late. Philosophy, as the thought of the world, does not appear until reality has completed its formative process, and made itself ready."[7] And yet, by openly acknowledging the implosion of an enduring tradition sustained by the combined and pervasive influence of Roman law and Christian theology—the "theologico-political machine" that has "marked for two millennia the Western conception of power" (*PC*, 18)—Esposito is also envisioning a performative role for philosophy: the deconstruction and retooling of political reason.

In order to accomplish this task, *Persons and Things* highlights the procedures of purification carried out by the axiological opposition of persons and things, showing how the intermediary of the body—in its manifold occurrences: flesh, animal body, political body, and biopolitical populations—presides over a vertiginous multiplication of splittings and hierarchizations. Persons and things are the juridical and theological operators of an anthropotechnical device of exclusion and subordination;

they reproduce within themselves the *summa divisio* of Roman law's *res/persona* and Christian theology's flesh/spirit.

The juridical person is a forensic term, suspended between the legal mask and the living body, the fictional attribution of responsibilities to individuals and their biological constitution. From Locke to Kant to contemporary neoliberalism, the concept of subjectivity reproduces in various ways the *res/persona* divide, duplicating itself into a *homo noumenon* and a *homo phaenomenon*, a rational part and a reified substance, the interiority of free subjectivity and the exteriority of natural necessity. A similar implosion takes place within the *res*, immediately divided by Roman law between the spiritual and the natural, *res corporales* and *res incorporales*, bodily and mental things, human and divine objects. In the realm of language, words progressively abandon their connection with things, introducing an alienation that devitalizes things and attributes meaning and truth to subjective consciousness. Marxism and psychoanalysis sustain this tendency, building their lexicon on a dialectic of the real and the symbolic, use and exchange value.

And yet, Esposito argues, the symmetrical polarity of bodies and things has been successfully performing its alienating operations, and the *res/persona* divide has been constantly reifying persons while desymbolizing things, subjugating an inferior part extracted from things and persons, thanks to the extra-juridical positioning of the body. Placed in an ambiguous and fluid middle ground, not thematized and de-autonomized, the body has guaranteed for centuries the translation and transition of persons into things, oscillating between them and allowing for the proliferation of variable thresholds. While Roman legal thought has programmatically cancelled out the body—which has no specific room in a clear-cut dualism of free persons and disposable things—philosophy has included it in a subordinated position. In both instances, the body allows for the coexistence of ideas and matter, *res cogitans* and *res extensa*, spirit and organism.

After denouncing the dualistic constitution of Western thought and the obscure role played by the body, Esposito shows how, already from within the philosophical tradition inaugurated by Plato and Aristotle, another conceptual lineage has been challenging the *res/persona* apparatus, overturning its premises and placing the body at the center. Reacting against the hegemony of philosophical modernity and its dualistic premises, Spinoza's one substance, Vico's historicism, Nietzsche's naturalism, Bergson's vitalism, and Husserl's and Merleau-Ponty's phenomenologies, have provided the most powerful alternatives, nurturing more recent exceptions to the theologico-political machine, such as Gilbert Simondon's technological vitalism and Bruno Latour's ethnographical approach to science studies.

Esposito suggests mobilizing these heretical traditions of Western philosophy, which have systematically opposed the repression of the body and challenged its subordination to the *res/persona* machinations. Since the regime of sovereignty has been replaced by the medical, demographical, and economic government of the biological life of populations, the human body must become also the active source of theoretical and political resistance.

Nonmodern Eyes

Esposito's most distinctive gesture is the unapologetic embrace of life and nature as positive territories for rethinking the political. Since traditional political categories, such as sovereignty and representation, are becoming exhausted simulacra, philosophy must think through the current biopolitical constellation, inside and against the government of life. As argued in *Bios*, biopolitics is not just the thanatopolitics of mass-murdering states and the politics over life of neoliberal democracies: it can as well become a politics of life, an affirmative politics of vitality.[8] Non-immunitarian communities, impersonal political agencies, and naturalistic practices are the building blocks of Esposito's reformulation of democracy in the biopolitical age.

In order to counteract modernity's negative biologizations of politics, *Living Thought* and *Communitas* appealed to Renaissance vitalisms and non-identitarian politics of friendship. *Persons and Things* attempts to recover nonmodern relations with things, as in ancient Roman law and Maori animistic rituals. Esposito's historicistic tendencies are thus superseded by a constituent archaism that reformulates the humanistic *topos* of the actuality of the past into a nonmodern destitution of linear historical time.

Thinking through Nietzsche and Benjamin, Mauss and Latour, Simondon and Sloterdjik, *Persons and Things* thematizes the "archaic and postmodern encounter of persons that are not persons anymore with things that are not things anymore" (*PC*, 102). On the one side, the transindividual territory of the body guarantees the spatial condition of possibility for a new alliance between things and persons, nature and history, science and politics; on the other side, the contemporaneity of a premodern connection of subjects and things offers an alternative temporal vector: "this is a sagittal relation between origin and completion, the archaic and the actual [. . .] that forces the historian, and even more the philosopher, to look beyond the most visible threshold of discontinuity" (*PC*, 99).

Esposito's "sagittal relation" between chronological strata of history cuts through the fictional linearity of history, linking apparently unrelated phenomena that belong to non-contemporaneous times. His genealogical method, which projects the premodern onto the contemporary, functions as a non-historicist description of cultural history and must not be confused with mystical or eschatological approaches, such as Bloch's principle of hope, Benjamin's *jetztzeit* (now-time), or Agamben's virtualities. It corresponds to Latour's generalized "principle of symmetry,"[9] to a bracketing off of nature and society and programmatic centering of philosophical investigation in the "Middle Kingdom" of quasi-objects and quasi-subjects that proliferate through bodies and technological artifacts.

Esposito delinks the reemergence of the archaic from the revelation of transcendence and Judeo-Christian salvific undertones. As in Latour, the nonmodern plane of immanence of things reappears when the ontological dualisms of modernity dissipate, when the mythography of Nature and Society is replaced by a materialistic look on the plethora of sociotechnological networks. Esposito is thus exposing Western philosophical tradition to the same "Copernican counter-revolution" called for by Latour's "symmetrical anthropology:" "if anthropology [. . .] is to be able to go back and forth between moderns and nonmoderns, it must be symmetrical. [. . .] the anthropologist has to position himself at the median point where he can follow the attribution of both nonhuman and human properties. [. . .] So long as we were modern, it was impossible to occupy this central place from which the symmetry between Nature and Society becomes visible at last, because it did not exist!"[10]

Instead of the asymmetrical relation between nature and society typical of modern epistemologies, according to which scientific methodologies explain truth of nature and sociological approaches account for errors and cultural codifications, symmetrical anthropology sees both nature and society as the by-products of an intense work of ontological purification, as an artificial construction superimposed on a material plane of technical, social, and existential practices. From the point of observation of hybrid, impure quasi-objects, of mixed realities of subjective and material things, nature and society occupy symmetrical positions that explain nothing and instead need to be explained as the outcome of real mediations.

Once the ethnographer positions herself in this in-between territory, she suddenly witnesses the evaporation of all tenets of Western modernity: premoderns stop being opposed to moderns, since they both occupy the nonmodern ecology of variable sociotechnological arrangements; the extra-human nature of the moderns ceases to be alienated from the interiority of culture, since biotechnological assemblages of quasi-objects take the place of the signs-things separation; the archaism of premodern worlds, in which nature and society are confused and mismatched by totemic and animistic affinities, does not stand anymore against the rational present of scientistic reason, since both of them are finally grasped as *a posteriori* constructs, colonial projections onto material networks of hybrids.[11] As in Latour's anthropology—which is describing "the world as we now see it through nonmodern eyes,"[12] after the illusionistic theatre of modernity has performed its last act—Esposito's sagittal relation with historicity unleashes the nonmodernity of modern times, rediscovering the constituent role of bodies and technologies, restituting subjectivity and intentionality to the silent realm of things: "in Brahmanic culture, the thing speak[s] in first person [. . .] the place where the power of the thing is exercised, and before that it is metamorphosed into a person, is the body of individuals and communities, of which it becomes an internal component" (*PC*, 97).

What sets apart Latour's critique of modernity from Esposito's deconstruction of philosophical categories is the importance attributed by Latour to modern science and colonial relations of power. While Esposito, in keeping with the analyses of Simone Weil (see *DP*, 23), locates the divergence of persons and things and the foreclosure of the body in the originary matrix of Roman law and Christian theology, Latour targets the scientific naturalism and the colonial logic of modernity, accusing them of having produced the two Great Divides that have been ruling modernity since the early seventeenth century. The first divide is a rigid partition of the human from the nonhuman, of culture and society from nature: an Internal Divide—internal to Western thought—that has shaped the ontological dualisms of persons and things, values and things-in-themselves, transcendence and immancence, producing the extra-human nature of Western modernity, "a Nature that is transcendent, unknowable, inaccessible, exact, and simply true."[13] The second generative split of modernity is an External Divide, the colonial distinction between Us and Them: on one side the West, as the locus where nature and nonhumans are distinguished from culture and humanity; on the other the premodern Others, the savages and barbarians of inferior civilizations, unable to distinguish signs from things, nature from culture.

According to Latour, the External Divide is a direct consequence, a simple export of the internal divide: "[I]f Westeners had been content with trading and conquering, looting and dominating, they would not distinguish themselves radically from other tradespeople and conquerors. But no, they invented science, an activity totally distinct from conquest and trade, politics and morality."[14] Coloniality and modern science are the two halves of modernity's arrangement of savagery and civilization, epistemic naturalism and culturalism. And yet, for Latour, it is the modern European separation of natural power from political power, it is the *Weltanschauung* that from Hobbes and Boyle has separated the science of things from the politics of men, the Laws of Nature from the Leviathan, a pure natural mechanism from a pure social force,[15] which can shed light on the colonial experience, explaining why non-Western modes of life have become, retrospectively, a premodern anthropological repertoire of scandalous mixtures.[16]

For both Latour and Eposito the asymmetry between nature and culture is a phantasmagoria that can be dispelled by a critique of Western ontologies and epistemologies. They also share the belief that when the bodily mixtures of things regain their protagonism, dissolving the nature/society divide, the historical relation of past and present gives way to a non-historicist composition of archaism and actuality, of animated things and impersonal subjects. A major divergence concerns their account of the origin of the divide of subjectivity and nature, placed by Esposito at the dawn of Roman history and by Latour at the onset of modernity. Because of these heterogeneous presuppositions, Esposito undertakes a critique of political theology that bypasses the critique of modernity, and Latour embraces an epistemological critique of modernity that excludes the postcolonial gaze.

States of Nature

If the metamorphosis of things and bodies, in our biopolitical and technoscientific age, is reshaping the old partition of subjects and matter; if the divide of science and society is being unmasked by decolonial thought as a component of a larger colonial geography of reason, then the architectonic balance of scientific naturalisms, social contract rationality, and extra-European savagery must crumble as well, together with the state of nature of modern political philosophy, the imaginary civilizational apparatus devised by Hobbes and Locke, Rousseau and Hegel. Because the two Great Divides are linked to each other, the critique of coloniality is now accompanying the critique of technoscientific rationality and political theology. Taken together, these two trajectories are questioning the state of nature of Western modernity: "Since politics has always been conducted under the auspices of nature, we have never left the state of nature."[17] The epistemic naturalism of the moderns, from Hobbes's mechanism to cognitive sciences' positivism, has doubled itself into the transcendentalisms and idealisms of the Euro-Atlantic philosophical tradition. Once nature loses its modeling function, becoming a hybrid terrain of mixed-reality and living things, a constructed matter and subjective aliveness, it is "no longer unified enough to provide a stabilizing pattern for the traumatic experience of humans living in society."[18]

The first consequence of this shift is that political philosophy and its founding categories—which, as in Hobbes's *Leviathan*, have been projected against the background of a mute, homogeneous parliament of things—lose the guarantee of their theatre of nature. Things become unsettling monsters, a nonmodern assembly of preoccupations and desires, demons that "interrupt any progression."[19] Nature, "instead of being a huge reservoir of forces and bottomless repository of waste" is now a pandemonium, populated by the specter of emancipated colonial savages and enigmatic quasi-objects.[20]

From within European philosophy, Martin Heidegger has sensed this shift, reacting to the collapse of the modern state of nature through a visionary archaism, deploying a powerful critical primitivism that has deconstructed modern scientific naturalism from the perspective of a premodern relation with things: "It is said that scientific knowledge is compelling. [. . .] Science's knowledge, which is compelling within its own sphere, the sphere of objects, already has annihilated things as things before the atomic bomb exploded. [. . .] The thingness of the thing remains concealed, forgotten. [. . .] the Old High German word *thing* means a gathering, and specifically a gathering to deliberate on a matter under discussion, a contested matter."[21] Heidegger focuses on this concealed "gathering" and develops an alternative definition of thingness, which questions the subject/object representational apparatus, the separation of things and persons, replacing it with a weaving of topological properties that complexifies things into a hybrid "thinging," at the intesection of "earth and sky, divinities and mortals."[22]

Like Esposito in *Persons and Things*, Latour reprises this Heideggerian intuition and expands the redefinition of thingness into a full-fledged nonmodern *Dingpolitik*, a "thingpolitics" that destitute the *Realpolitik* and body politics of modern political philosophy: "no doubt, the Body Politics is a monster—so much so that it's not even a body. But which type of monster is it?"[23] Once nature abandons the naturalism of modern scientific reason and transcendental critical philosophies, politics is confronted by the puzzling arrangements of heterogeneous, dense and immanent things, by uncanny assemblages of technosocial hybrids asking for new assemblies. Things of all kinds gather and pertain, concern and question. They are not the usual objects, a calculable matter of fact, but "bodies without organs," unstable matters, automated or catatonic, endowed with demands and needs or empty and passive. "Scientific laboratories, technical institutions, marketplaces, churches and temples, financial trading rooms, Internet forums, ecological disputes"[24] are the quasi-subjects of a contemporary, nonmodern *Dingpolitik*.

The rebellion of bodies captured by Esposito's biopolitics and Latour's technoscientific networks is accompanied by the emergence of nonmodern paradigms repressed by Western "mono-naturalism."[25] The emancipation of thought from the subject/object, archaic/contemporary divides, the deconstruction of modernity's state of nature have unleashed nonmodern modes of ontological identification[26] and triggered the proliferation of age-defining neologisms: terms such as Anthropocene, Capitalocence, Chthulucene, and Gaia are attempts to apprehend states of life and nonlife, humanity and nonhumanity that reject the traditional framework of political theory.[27] This mostly unacknowledged, or disguised, return of nonmodern state of nature paradigms carries within itself also the return of the haunting figures of the sovereign and the beast, the contract and the savage, the socialized and the wild.[28] Postnatural states of nature are reshaping our theoretical, aesthetic, and political imagination.[29] The effort to account for the political agency of unfamiliar webs of life and nonlife, animation and animalization, contagion and immunity has profoundly modified the state of nature/state of society conceptual apparatus, without neutralizing its structural mechanisms.[30]

One instance of this redistribution of concepts is Félix Guattari's "machinic animism"[31] and its resonances in Eduardo Viveiros de Castro's and Philippe Descola's anthropologies. Guattari's machinic animism is the name for an extension of subjectivity beyond the limits of the Eurocentric dualisms of persons and things, the manifesto of an "artificial alliance between animism and materialism:"[32] "if I understand Guattari, the first thing to do is to cut off the relation between the subject and the human. Thus subjectivity is not a synonym of humanity. The subject is a thing, the human is another thing. The subject is an objective function that one can find deposited on the surface of everything. [. . .] That is how it is for Amazonians. For them, the subject is a way to describe the behavior and attitude of things, just as for us, objectivation is a way to describe things in this sense."[33]

Eduardo Kohn has pursued a similar project, embracing the posthuman critique of human exceptionalism and animistic and shamanic categories—along with Peirce's semiotics and Deleuzian-influenced ontologies—in order to "flatten the distinction between humans and other kinds of beings, as well as those between selves and objects."[34] Analogously, in *Vibrant Matter*, Jane Bennett has sketched out a "political analysis that can better account for the contributions of nonhuman quasi-subjects."[35] Bennett's assemblage of actor-network categories and vitalist ontologies recognizes the urgency to articulate a politics of the nonhuman, the need of coming to terms with the political implications of "active powers issuing from nonsubjects" and "circulating around and within human bodies,"[36] thus forcing political theory to rethink its nature, recognizing the challenges posed by material formations not centered on human subjectivity.

"Gaia theories," invoking the primal Earth goddess, the Greek Mother of Western gods, are also rediscovering the political actuality of animism. Departing from the original formulations of the Gaia hypothesis by James Lovelock and Lynn Margulis—who introduced Gaia as the figure of our "living planet," a description of the Earth as a vital, self-regulating cybernetic system with homeostatic tendencies—two significant, although divergent philosophical cults of Gaia have emerged in recent years: one introduced by Isabelle Stengers in her 2009 volume *In Catastrophic Times*; the other presented by Bruno Latour in several essays and most notably in his 2013 *Gifford Lectures* on "the political theology of nature."[37] The Rio de Janeiro 2014 conference "The Thousand Names of Gaia: From the Anthropocene to the Age of the Earth," organized by Eduardo Viveiros de Castro and Bruno Latour, has then imposed the convergence of such Gaia philosophies and the Anthropocenic vocabulary.[38]

Latour's declination of Gaia continues Michel Serres's programmatic return to Hobbes and ambition to redesign the social contract as a "natural contract." Latour pushes the natural contract of Serres even further, since for him, "in an unexpected and unprecedented twist on Hobbes's most famous concept" we have entered a planetary "new state of nature" and, in this situation, we need to "search for a new Leviathan that would come to grasp with Gaia."[39] Like Behemoth, the Biblical monster that frames Hobbes's history of the English civil war, Gaia must be tamed by a new Leviathan, since she is the goddess of our contemporary ecological state of nature, with its "war of all against all, in which the protagonists may now be not only wolf and sheep, but also tuna fish as well as CO_2, sea levels, plant nodules or algae, in addition to the many different factions of fighting humans."[40] Latour presents himself as the new Hobbes, the high priest of Gaia, the shaman of a menacing "theology of nature," with the goddess Gaia replacing the wild deities of Hobbes's savages, and "a new civilized collective"—the institutions assembled for administering the cult of Gaia—composing her anarchic power, prolonging Hobbes's artificial commonwealth, and coming to terms with the culture wars of the Anthropocene.[41]

Despite its epistemic naïvety and lack of political and cultural sensitivity, analytic philosophy has tried as well, since the publication of Thomas Nagel's seminal essay "What Is It Like to Be a Bat?"[42] to extend "mental properties" beyond the human, in the attempt to thematize the animacy of nonhuman beings. Current debates on new materialisms, speculative realisms, neo-vitalist transcendentalisms, and other object-oriented ontological paradigms have recuperated the term "panpsychism," suggesting the existence of "a long philosophical pedigree," a "recurring underground motif in the history of Western thought" that runs "from the pre-Socratics, on through Baruch Spinoza and Gottfried Wilhelm von Leibniz, and down to William James and Alfred North Whitehead," a *philosophia perennis* centered on the idea that "mind is a fundamental property of matter itself" and "thinking happens everywhere."[43]

In an age marked by a global ecosocial mutation—climate change, geological and environmental catastrophes, demographic shifts and genocides, mass pauperization, biopolitical and bioengineering techniques of government of the populations—philosophical categories are forced to rethink their presuppositions. The states of nature emerging from neo-animist and naturalistic modes of relation between humans and nonhumans are the signs of the nonmodern consciousness of our times. The puzzling life of "factish" things is disclosing a lifeworld in which shamanic rituals and technosciences, political ecologies and indigenous mythograms cohabit.[44] This is neither a peaceful world, nor the Eurocentric phantasy of a global state of society.

Notes

1. Martin Heidegger, "The Thing," in *Poetry, Language, Thought*, trans. Albert Hofstadter (New York: Harper Collins, 1971), 168.

2. See Esposito's *Categories of the Impolitical*; *Communitas*; *Bios*; *Immunitas*; *Third Person*; and *Terms of the Political*.

3. See Esposito's *Living Thought*; *Two*; and *Da fuori* (English translation *A Philosophy for Europe: From the Outside*, trans. Zakiya Hanafi, forthcoming at Polity Press).

4. All the English translations from *Le persone e le cose* in this paper are mine.

5. Dipesh Chakrabarty, *Provincializing Europe: Postcolonial Thought and Historical Difference* (Princeton, NJ: Princeton University Press, 2007).

6. In Viveiros de Castro's words, "there are more subjects than humans. Subjectivity is a fusion of multiplicity, not of unity. It produces not a unity of consciousness or a function of integration. It is a function of dispersion [. . .] This is animism, the idea that the subject is outside. It is everywhere" (Interview with Eduardo Viveiros de Castro, in Angela Melitopoulos and Maurizio Lazzarato, "Assemblages: Félix Guattari and Machinic Animism," *e-flux* 36 (July 2012): 4, 7).

7. Georg Wilhelm Friedrich Hegel, "Preface," in *Philosophy of Right* (Mineola, NY: Dover Publications, 2012), xxi.

8. See Timothy Campbell, "Bios, Immunity, Life: The Thought of Roberto Esposito," *diacritics* 36, no. 2 (2006): 2–22.

9. Bruno Latour, *We Have Never Been Modern*, trans. Catherine Porter (Cambridge, MA: Harvard University Press, 1993), 94.

10. Ibid., 91, 96.

11. "Real as Nature, narrated as Discourse, collective as Society, existential as Being: such are the quasi-objects that the moderns have caused to proliferate" (Latour, *We Have Never Been Modern*, 90). Among the quasi-objects mentioned by Latour are "frozen embryos, expert systems, digital machines, sensor-equipped robots, hybrid corn, data banks, psychotropic drugs, whales outfitted with radar sounding devices, gene synthesizers, audience analyzers" (Ibid., 49).

12. Latour, *We Have Never Been Modern*, 103.

13. Latour, *We Have Never Been Modern*, 90. See also Willard Van Orman Quine, "Epistemology Naturalized," in *Ontological Relativity and Other Essays* (New York: Columbia University Press, 1969).

14. Latour, *We Have Never Been Modern*, 97.

15. Ibid., 30.

16. For an opposite perspective see Anibal Quijano, "Coloniality of Power, Eurocentrism, and Latin America," *Nepantla* 1, no. 3 (2000): 544. "The Eurocentric pretension to be the exclusive producer and protagonist of modernity—because of which all modernization of non-European populations, is, therefore, a Europeanization—is an ethnocentric pretension and, in the long run, provincial." For a more complex diagnosis of the relations of science and coloniality, see the special issue on "Science, Colonialism, Postcolonialism" of the journal *Postcolonial Studies* 12, no. 4 (2009).

17. Bruno Latour, *Politics of Nature: How to Bring the Sciences into Democracy*, trans. Catherine Porter (Cambridge, MA: Harvard University Press, 2004), 235. With a memorable dictum, the anthropologist Eduardo Viveiros de Castro has identified the task of contemporary thought: "[T]there is a long tradition in the history of philosophy in which imaginary savages were used in arguments within real philosophy. By way of reciprocation, I will claim the right to summon real savages to help me to do a bit of imaginary philosophy" (unpublished paper, *The Other Metaphysics and the Metaphysics of Others*, 2012).

18. Bruno Latour, "From Realpolitik to Dingpolitik—An Introduction to Making Things Public," in Bruno Latour and Peter Weibel, eds., *Making Things Public: Atmospheres of Democracy* (Cambridge, MA: MIT Press, 2005), 29.

19. Latour, "From Realpolitik to Dingpolitik," 30.

20. Ibid., 15.

21. Martin Heidegger, "The Thing," 168, 172.

22. Ibid., 176.

23. Latour, "From Realpolitik to Dingpolitik," 28.

24. Ibid., 22.

25. See Eduardo Viveiros de Castro, "Cosmological deixis and Amerindian perspectivism," trans. Elizabeth Ewart, *Journal of the Royal Anthropological Institute* 4, no. 3 (1998): 469–88; and Philippe Descola, *Beyond Nature and Culture* (Chicago: University of Chicago Press, 2013), 172–74.

26. On the return of the political figure of the Animist, see Elizabeth A. Povinelli, *Geontologies: A Requiem to Late Liberalism* (Durham, NC: Duke University Press, 2016). On the political implications of nonhuman forms of life and vitality see also Mel Y. Chen, *Animacies: Biopolitics, Racial Mattering, and Queer Affect* (Durham, NC: Duke University Press, 2012).

27. Latour's observation that in our time, "contrary to Hobbes's scheme, the 'state of nature' seems to have a dangerous tendency to *follow*, and not to precede

or to accompany, the time of the civil compact" may be correct (Bruno Latour, "Agency at the time of the Anthropocene," *New Literary History* 45, no. 1 (2014): 6). Latour, as Michel Serres in *The Natural Contract*, trans. by E. MacArthur and W. Paulson (Ann Arbor: University of Michigan Press, 1995), endorses Hobbes's lexicon and the nature-society dualism, which is shaping his Gaia theory. Latour describes himself as a "new Hobbes," the demiurge of a new Leviathan governing an anthropocenic state of nature, the state of war produced by climate change and contemporary technosciences; see Bruno Latour, *Face à Gaïa. Huit conférences sur le nouveau régime climatique* (Paris: Éditions La Découverte, 2015).

28. See for instance Jodi A. Byrd's analysis of the "political bestiary of sovereignty" and state of nature "pathology" of political philosophy, and their implications for indigenous politics: "That the Indian is the foundational pathology within the logics of modern sovereignty is not a given within Western thought, though clearly the juxtaposition of 'the Savage as the Wolf,' and the establishment of the line of exception/unexception that marked the limits of U.S. settlement, signals the role the bestial Indian played as the dialectical bind that constituted U.S. sovereign civility at the beginning of its formation as a nation-state" (Jodi A. Byrd, "Mind the Gap," in F. Luisetti, J. Pickles, and W. Kaiser, eds., *The Anomie of the Earth: Philosophy, Politics, and Autonomy in Europe and the Americas* (Durham, NC: Duke University Press, 2015), 127).

29. For an interpretation of Foucault's biopolitics as a state of nature epistemology see Federico Luisetti, "Notes on the Biopolitical State of Nature," *Paragraph: A Journal of Modern Critical Theory* 39, no. 1 (2016): 108–21.

30. Examples of the pervasiveness of the state of nature apparatus are Giorgio Agamben's *homo sacer* political ontology—in which the biopolitical dialectic of *zoe/bios* substitutes the binary of the state of nature/society (Giorgio Agamben, *Homo Sacer: Sovereign Power and Bare Life*, trans. Daniel Heller-Roazen (Stanford, CA: Stanford University Press, 1998) and *The Open: Man and Animal*, trans. K. Attell (Stanford, CA: Stanford University Press, 2004)), as well as Jacques Derrida's explorations of the dialectic of sovereignty and animality (Jacques Derrida, *The Beast and the Sovereign*, Volumes 1 and 2 (Chicago: University of Chicago Press, 2011).

31. See Angela Melitopoulos and Maurizio Lazzarato, "Machinic Animism," *Deleuze Studies* 6, no. 2 (2012): 240–49.

32. Eduardo Viveiros de Castro in Melitopoulos and Lazzarato, "Machinic Animism," 242.

33. Interview with Eduardo Viveiros de Castro in Melitopoulos and Lazzarato, "Assemblages," 4.

34. Eduardo Kohn, *How Forests Think: Toward an Anthropology Beyond the Human* (Los Angeles, CA: University of California Press, 2013), 7. For a "multispecies" neo-animist paradigm, see Donna Haraway, *Staying with the Trouble: Making Kin in the Chthulucene* (Durham, NC: Duke University Press, 2016).

35. Jane Bennett, *Vibrant Matter: A Political Ecology of Things* (Durham, NC: Duke University Press, 2011), x.

36. Ibid., ix.

37. Isabelle Stengers, *In Catastrophic Times: Resisting the Coming Barbarism*, trans. Andrew Goffey (Open Humanities Press, 2015); Bruno Latour, *Facing Gaia: Six Lectures on the Political Theology of Nature. Gifford Lectures on Natural Religion* (Edinburgh, February 18–28, 2013), now published in French as *Face à Gaïa. Huit conférences sur le nouveau régime climatique* (Paris: Éditions La Découverte, 2015).

38. See https://thethousandnamesofgaia.wordpress.com/

39. Latour, *Facing Gaia*, 103–04.

40. Ibid., 103.

41. According to Latour, "just as Hobbes needed the state of nature to get to the social contract, we might need to accept a new state of war to envision the State of peace" (Latour, *Facing Gaia*, 114).

42. Thomas Nagel, "What Is It Like to Be a Bat?" *Philosophical Review* 83, no. 4 (1974): 435–50.

43. Steven Shaviro, "Consequences of Panpsychism," in Richard Grusin, ed. *The Nonhuman Turn* (Minneapolis, MN: University of Minnesota Press, 2015), 20.

44. See Bruno Latour, *On the Modern Cult of the Factish Gods* (Durham, NC: Duke University Press, 2010). On the political ecologies of social movements and decolonial relations to nature, see Arturo Escobar, *Territories of Difference: Place, Movements, Life, Redes* (Durham, NC: Duke University Press, 2009) and Walter D. Mignolo, *Local Histories/Global Designs: Coloniality, Subaltern Knowledges, and Border thinking* (Princeton, NJ: Princeton University Press, 2012).

Part II

APPLYING ESPOSITO

9

The Impolitical Dimension in Jorge Luis Borges' Literature

A Gaze on the Impossible for Politics through Roberto Esposito's Thought

Federico Fridman

Introduction

A fundamental conceptual step in Roberto Esposito's intellectual trajectory toward his more recent work on biopolitics and on the concept of the impersonal was his book *Categories of the Impolitical*, which was first published in 1988 and re-edited in 1999 with a new preface.[1] By the end of the 1980s, facing the appalling domination of the neoliberal model and the failure of political struggles for political emancipation, Esposito performs a radical torsion on crucial categories of political thought in order to keep thinking about politics under rather discouraging circumstances. He elaborates the notion of the "impolitical" ("*l'impolitico*" in Italian), which he derives from Massimo Cacciari's influential essay "L'impolitico nietzscheano," as a lens through which to offer an incisive view of modern politics.[2] The impolitical crystallizes a complex displacement in Esposito's philosophy, which withdraws from a representational form of thought that pretends to either represent political conflict or to organize it. Esposito's impolitical view perceives an irreducible conflict at the heart of the political that can only be represented through a violent reduction of its intrinsic plurality. Thus, he retreats from all forms of political philosophy that subordinate one term under another, whether imposing a philosophy on the part of the political or directing the political according to philosophical principles.[3]

Esposito's elaboration of the impolitical remains somewhat obscure because he does not provide a definition of this elusive concept. Although the title of the book is *Categories of the Impolitical*, he prefers to view the impolitical not as a category but rather as a way of looking. He asserts that the impolitical is a question of tonality, of nuance, a mode of seeing the political.[4] Certainly another reason why the impolitical is difficult to comprehend is because it emerges in opposition to representation. In the opening section of my paper, entitled "The Impolitical Gaze," I will shed light on the impolitical form of thought by addressing its constitutive opposition to an anti-political position. Next, I will explain the anti-theological political attitude that characterizes the impolitical. I then explore through an impolitical perspective the particular relationship that modernity articulates between the political and *tekhnè* to design new mechanisms of domination, and the consequences of these new configurations for the individual. This section closes with an analysis of the impolitical radical critique of form to represent politics.

In the section entitled "The Impolitical Dimension in Borges's Literature" my intention is to show how the impolitical comes into play in Jorge Luis Borges's texts. The conception of the impolitical in Italy relies utterly on a Eurocentric perspective. Esposito tacitly develops his retrospective view of the social and political processes that led to totalitarian regimes in Europe and offers a radical critique of contemporary democratic regimes, which have been unable to provide an alternative to power and domination. Nevertheless, on the other side of the Atlantic divide, beginning in the 1930s, Borges cunningly sees through an impolitical lens the imminent collapse of Europe under the rising shadows of European totalitarian regimes. His literary and intellectual productions do not cease to map, interpret, and rethink the consequences of these events from an impolitical refractory angle, while integrating his alarming perception that democratic social and political processes in the Argentinean context may follow in a similar direction. Borges's writing configures an impolitical dimension that not only provides an alternative way to grasp Esposito's impolitical form of thought, but also offers readers the opportunity to analyze from an innovative perspective his strong disagreement with the political as a representation of the social.[5]

In the last section "Ground Zero: The Social and Political," I bring to light how Borges's and Esposito's impolitical attitudes converge. Nevertheless, I then examine the ways that Borges's and Esposito's answers diverge regarding the political. I point out Borges's ideological position against the political, as well as the paths that Esposito follows, after *Categories of the Impolitical*, to locate the possible reconfiguration of the political mechanisms that capture and control biological human life.

The Impolitical Gaze

The impolitical is not a loss of interest in politics, or a weakening of politics, but rather the intensification and radicalization of the political and its limits: the impolitical is not anti-political. Esposito writes: "[T]he anti-political cannot coincide with the impolitical because it already coincides with the political."[6] The anti-political does not deny the political but rather presents itself as an inverted image of the political, a negative reflection. The anti-political is a way to engage the political by opposing it; hence, it reproduces what defines the political: enmity and opposition. Therefore, the constitution of the anti-political is still political because it adopts the essential form that characterizes the political. The anti-political denies the original conflict that justifies the political. However, as the anti-political depoliticizes this conflict while postulating a representation of such neutralization, it reproduces the political structure of representation and, by inverting this image, draws a line of confrontation against it.

The impolitical, on the other hand, inhabits the political. Esposito writes: "[. . .] the impolitical coincides with the political precisely because it does *not* negate it."[7] The impolitical view perceives that there is no other power than the political itself, and that this force cannot be opposed by another force with a different nature. The potentiality of the political is limited by what it is. Hence, instead of colluding with the political and trying to organize it, instead of denying that the political is an irreducible conflict, the impolitical gaze regards the political as the only reality. However, it is only reality. This way of looking, which reminds us of the finitude of the political, intensifies the political by stretching it to its limits, and once these edges have been reached, it turns 180 degrees and sheds light on what remains impossible for the political: combining good and power.

Adopting Esposito's view, we might say here that a theological understanding of the political relies on "the conception according to which the good would be politically representable and politics would be interpretable in terms of value."[8] The impolitical gaze points out precisely an irremediable rupture between good and power. Esposito writes: "[P]ower is neither a representation nor an emanation from it, and still less a dialectical mechanism capable of recuperating Good from Evil, thus converting Evil into Good."[9] The transition that modernity provokes in language from theological to juridicopolitical concepts, which coincides with the process of secularization, rather than drying out the transcendental source that legitimizes political decisions instead creates a re-enchantment of the political. The impolitical view recognizes that in the modern desacralizing of power there is the formation of a normalizing apparatus that is inevitably enveloped in sacred attributes. Thus, Esposito states that his work begins precisely where Carl Schmitt's *The Concept of the Political* ends.

A complex constellation of authors converges in *Categories of the Impolitical*, but I would like to particularly draw attention to Esposito's disclaimer regarding Schmitt's work, because this represents a fundamental conceptual displacement in his philosophy.[10] Bruno Bosteels argues that the impolitical attitude, following Nancy's and Lacoue-Labarthe's logic of the "withdrawal of the political," is a matter of avoiding the necessity of drawing a line of confrontation and assuming a partisan commitment; or, in other words, of suspending the Schmittian political distinction between friend and enemy.[11] Nonetheless, the impolitical form of thought does not deny that this confrontation is constitutive of the political; rather, it rejects the idea that there can be a positive representation of this conflict—or even better, it claims that any political representation of such a conflict is inevitably invested with a transcendental element. Therefore, the impolitical opposes Schmitt's work because his genealogical theory of sovereignty coincides with the same secularized transition from theological language to the juridicopolitical. And although Schmitt points out an original substantial emptiness in the political decision, he seeks a functional transformation of the theological definition of sovereignty.[12] Schmitt does not fracture the political-theological structure that legitimizes political decision making, but rather he invests the political with a gravity similar to that of a sovereign's decision before modernity, when it was legitimized by a theological and transcendental element.

Esposito's analysis focuses first on Schmitt's book *Roman Catholicism and Political Form* to begin the process of deconstructing the political-theological structure that exists untouched beneath the secularized structure of modern states. In this text, Schmitt posits that depoliticizing political representation is an intrinsic characteristic of modernity. He asserts that the rejection of a form of representation that bonds the political decision to the idea or, in other words, that consists of a positive relationship between the good and power, constitutes a crucial feature of modernity. Schmitt writes: "Once the state becomes a Leviathan, it disappears from the world of representations."[13] Nevertheless, he does not ignore the fact that Thomas Hobbes initiated the history of modern political representation by founding the relationship between the sovereign and his/her subjects on a plane of immanence. Schmitt points out that since the sublime content that granted to this relation a transcendental otherness has been leveled, the political has been degraded. Modernity cuts the vertical thread that legitimized the political. This is a radical negation of the Hobbesian idea because there is no longer a pure image to which the political decision can refer for its foundation and its purpose.

The Catholic *repraesentatio* reacts against the rupture caused by modernity in an attempt to stitch together the broken link between good and power. Esposito points out that Romano Guardini, in line with this response, claims that the justification for a political decision cannot be either framed by or emanate from legal norms, but that it must incorporate the representation of a transcendental substance.[14] The immanent and the transcendent poles must remain intimately linked together and converge in the figure of the sovereign. The impolitical form of thought undermines this bridge. Esposito argues that it absolutely separates "[. . .] men and god, nature and divine grace, time and eternity" (*CIM*, 32).[15] Nevertheless, he points out that this attitude does not aim to dismantle the Catholic reaction on the basis of a definitive affirmation of modern political representation. On the contrary, the "[i]mpolitical is precisely that attitude [. . .] that even while rejecting the depoliticizing success of modern secularization, and rather situating itself at its antipodes [. . .] rejects at the same time all falling back on any theologicopolitical *repraesentatio*, any transcendent place for *grounding* the political."[16]

In modernity, the theological political posture attains in its secularized facets its most effective modality, which is not, as Schmitt believes, an excessive appreciation of the modern form of political representation.[17] The transition from a theological to a secular legitimization of the political organizes the original emptiness—the lack of intrinsic unity—that dwells within the political. The modern democratic state—liberal, agnostic, and neutral—can only produce a synthesis, as it occupies the role of mediator between various economic interests. However, it does not fill the emptiness left by the lack of a transcendental element; instead, it organizes this emptiness under its authority. It is the guarantor of the autonomy of the economy so that the individual can exercise his/her freedom to exchange goods in the marketplace. Thus, according to Esposito, the profound depoliticization enacted by the modern State reveals a serious contradiction because the autonomy of the economic sphere restores the *hyper-political* character to the political. Paradoxically, this modern depoliticization involves a highly political effort that is necessary to establish and preserve the conditions that permit the autonomous functioning of the market.

The secular State disperses power when it creates a sphere in which the power of the political union is fragmented into different elements motivated by economic interests. However, this dispersion itself must in turn be represented. Lacking a foundation that legitimizes the political relationship between the sovereign and his/her constituency, this relationship must find a new source of legitimacy, one which can only be a myth. Consequently, the modern desacralizing of power provokes a second theologization of political representation. The impolitical attitude challenges the myth of the coexistence in modernity of the depoliticized theology; it does so by refusing to assign any other value to politics and thus it breaks politics' spell. There is no value to be opposed against the political, but rather the contrary. There is only the negation of the political as a value, the negation of a theological assessment of its value, and the negation of the positive representation of power.

Politics is an unrepresentable plurality because its origin is always plural and outside of any representative source. Esposito writes: "[E]very logic-historical attempt to represent such plurality in fact constitutes an evident negation of it, since the intrinsic modality of representation is the *reductio ad unum*" (*CIM*, 15).[18] In order to make political representation possible it is necessary to identify a third imaginary element that transcends the relationship between a representative sovereign and his/her constituency. This third element is the idea that positively defines the relationship through the transcendental effect of a Platonic idealism. *Reductio ad unum*—people, nation, State—can only happen if representation is theological at its top, through the divinization of sovereignty, and depoliticization at its base. Yet multiplicity remains unrepresentable. This is the reason that a third element, capable of imbuing political representation with a transcendental and therefore imaginary meaning, not a substantial one, is necessary. This need for self-legitimization of the political explains why this relation is always linked to a theological model, "and hence fatally connected to its theatrical representation" (*CIM*, 16).

The impolitical gaze understands that the origin of politics is a non-origin; it is a continuum of chaos and plurality that is always present and contemporary. Thus, it cannot be represented or reduced to unitary symbols. And there has never existed a natural order or a unitary symbol to represent this order that was subsequently profaned by technology. There is a tradition of authors, including Max Weber, Schmitt, and Arendt (in her less impolitical facets), who have concluded that contemporary politics has betrayed itself and its proper essence, falling sway to the power of *tekhnè*. However, the impolitical deconstructs precisely this perspective. There is neither an *arché* nor an *ousia* of politics. *Tekhnè* is not the end of politics, because there has not been a transition from politics to *tekhnè*. The political originated with and within *tekhnè*; it is its origin and, at the same time, it is a non-origin. Esposito writes: "[T]he impolitical recognizes the perfect co-originariness of technology and politics; recognizing also that [. . .] there does not exist a praxis that is prior to *tekhnè* or qualitatively different from it."[19]

The distinctive characteristic of modernity is that it has engendered a previously unknown tangle between *tekhnè* and the political to underpin subjugation. The transition from feudalism to modernity set in motion a process that depoliticized civil society through the configuration of different centers of neutralization. The *tekhnè* that is used in these centers, as pure political neutralization, is the main force that controls them. This neutrality and its wide availability—the possibility for this new *tekhnè* to be used by any political force that has the power to dominate it—has made it both the point of departure for depoliticization and an absolutely apolitical element. However, Esposito claims that due to this neutrality *tekhnè* "becomes the object of an extreme politization, and even prevents subsequent despoliticizations" (*CIM*, 57). The depoliticized essence of *tekhnè*, which makes it possible for it to be appropriated by any form of power, leads only to a more radicalized politicization.

Esposito asserts that the manipulations of *tekhnè* provoked a qualitative leap in modern politics. On the one hand, the relationship between the individual's subjectivity and function was reconfigured. On the other, the relationships between knowledge, action, and experience were also changed. "Whereas in the modern age *tekhnè* still alluded to a subject-man (the *homo faber*) considered as the immobile fundament of his cognitive activity and practice, nowadays the new *tekhnè* precisely raises from the irreversible collapse of the *homo faber*" (*CIM*, 59). Thus, *tekhnè* has a particular relationship with the masses, which can no longer be understood as the sum of individual entities, but rather is structured by a different normalizing principle. The redefinition of the status of a man/woman as such, a status no longer possible to define in terms of a subject that is the fundament for his/her own activity and practice (subject-fundament), is the result of a radical transformation of subjectivity that creates a gap between knowledge, action, and experience, and is therefore a rearrangement of the political arena.

The impolitical form of thought assumes that there is neither an emancipatory origin nor an essence of politics that can be rescued. In a dialogue with Nancy, Esposito says: "[. . .] there can be no doubt that every road back is blocked; every return to the natural origin [of politics] is impossible since, as we have seen, the origin does not exist per se."[20] The impossibility of representing the origin of politics opens a space for the impolitical. Esposito explains: "The impolitical is nothing other than the enunciation of this unrepresentability. It is the 'withdrawal' [*ritratto*]—in the sense of 'to withdraw itself' [*ritirarsi*], and in that of 'to re-mark' [*rimarcare*]—of originary trace."[21] The impolitical withdrawing from all attempts to determine the origin of politics fundamentally points out the impossibility of all forms of political representation and, furthermore, offers a radical critique of form in order to represent the irreducible plurality that is forced to converge under the political.

How then can we configure an impolitical form of writing, which, while rejecting representation, is also capable of withdrawing and folding in on itself to enunciate the absolute negativity that defines the political? Esposito's writing relies on the lexicon of the tradition of Western political thought to advance into the theoretical displacements that he analyzes, which draw the contours of the impolitical dimension. Nevertheless, as mentioned before, he refuses to consider the impolitical a category or a concept, but rather a tonality, a nuance, a way of looking that remains impossible for the political, as if theoretical language has reached an impasse in trying to grasp the ground zero that dwells within the political. Hence, would the realm of literature be precisely a territory in which the tonalities and nuances of the impolitical perspective could achieve more intensity and clarity? Is there an intrinsic aesthetic in the mode of existence of the political that the impolitical gaze is able to capture and that precisely a literary way of thinking and writing could communicate?

The Impolitical Dimension in Borges's Literature

After the 1930s, Borges opens an impolitical dimension in his texts in which he echoes an attitude that Franz Kafka and Macedonio Fernández, two writers who decisively influenced his work, assumed regarding the political. Under the consolidation of the modern democratic liberal state and the rising shadow of European totalitarian regimes, these writers perceived in the new forms of the political an imminent threat that today remains enshrouded in darkness: modern politics cannot provide an alternative to oppression. Kafka and Fernández radicalize the political mechanisms of domination, discipline, and control in their texts to identify the new forms of power already knocking at their doors. Borges particularly will extend this perspective to reflect not only on politics, but also on social and political struggles that aim to achieve the power of the state, and on the modern political party. He will express a profound disagreement with contemporary politics as the representation of the social: a form of thought that radically rejects the political as a means to achieve emancipation.[22]

In the short story "El atroz redentor Lazarus Morell" ["The Dread Redeemer Lazarus Morell"] written in 1935, Borges suggests a tragic destiny for modern politics and a lethal trap in political representation. In the story, which takes place at the beginning of the nineteenth century in the vast cotton fields along the Mississippi River, Borges draws an analogy between the plan sketched by the main character, Lazarus Morell, to subjugate others for profit and the mechanisms of domination and cooptation that were configured in the first decades of the twentieth century by charismatic leaders and political parties. The narrator says that Morell planned out the ploy that would assure him his rightful place in the universal history of infamy while he was captivating an audience from the pulpit preaching Scripture and his men were outside stealing the listeners' horses, which were then sold in another state.

Morell developed a unique modus operandi that is distinguished by the sordidness it required and by its deadly manipulation of hope. He commanded a thousand men who had sworn loyalty to him. These men traveled all around the vast plantation lands of the American South. The narrator explains: "They would pick out a wretched black and offer him freedom. They would tell him that if he ran away from his master and allowed them to sell him, he would receive a portion of the money paid for him, and they would then help him escape again, this second time sending him to a free state."[23] This procedure would be repeated more than once, and they would tell him that they had extra expenses and thus needed to sell him again. "On his return, they said, they would give him part of both sales and his freedom. The man let himself be sold [. . .]."[24] After he had escaped for a third time, desperately seeking freedom, the gang members would murder him and dispose of his body in the river.

Lacking any news about the slave who had escaped, the other slaves thought that he had been released and thus they believed that Morell was their redeemer. However, he was not an emancipator. He knew how to manipulate the slaves' hope in order to one day have his own cotton plantation with his "rows of bent-over slaves."[25] The operation grew and it became necessary to take on new *afiliados* [members], until one of the new gang members broke his oath and exposed Morell and the rest of his men. Although he was chased and surrounded by authorities, he succeeded in evading them and tried to put together a personal army. Finally, Morell died of a lung ailment in a hospital in the city of Natchez.

The diabolical nature of Morell's scheme resides not only in the fact that he was not an emancipator, but also in the fact that the subjugation that was repeatedly imposed was essentially inflicted on the slave by himself because he believed that a state of freedom was possible to achieve. It is thought provoking to consider that Borges covertly suggests an analogy between Morell, the preacher and "redeemer" of slaves, and the charismatic political leaders who attained the power of the state with the support of mass-based political parties in the mid-1930s.[26] There is a specific trace in his texts that I believe invites us to consider this possibility. In alluding to the new members of the gang, the narrator calls them *nuevos afiliados* [new members]. In Spanish, the word *afiliados* is often used to refer to the members of a political party. Thus, it is interesting to consider that Borges offers a radical critique not only of the charismatic leader who manipulates the hope of the masses promising that he will lead them to a state of freedom, but also of the political structure of the party that coopts new *afiliados*, as if they were gang members who will participate in a vast, evil machination.

Borges rejects any possibility of grandeur in modern politics. He believes that the political is based on the art of simulation. As he claims in "El pudor de la historia" ["The Modesty of History"] written in 1952, spectacular historical political events are so numerous because "one of the tasks of governments (especially in Italy, Germany, and Russia) has been to fabricate them or to simulate them with an abundance of preconditioning propaganda followed by relentless publicity."[27] In other texts, Borges particularly observes the similarities between the propaganda techniques used by totalitarian regimes in Europe and the theatrical propaganda techniques that Juan Domingo Perón's regime (1946–1955) employed in Argentina. It was irrelevant for Borges that Perón was democratically elected or that the military coup d'état that overthrew him in 1955 established a dictatorship.

In "L'illusion comique," for instance, a short essay that Borges wrote a few months after the coup in 1955, he points out how political power manipulates specific forms of *tekhnè,* using propaganda and theatrical techniques to decisively influence the people's process of *subjectivation.* Borges first addresses "the methods of commercial advertising and of *littérature pour concierges* [that] were applied [by the Peronist regime] to the governing of the Republic."[28] He then claims that consequently there were two histories, one of a criminal nature and the other of a theatrical nature: "tales and fables made for consumption by dolts."[29] Borges asserts that this latter history compelled him to write his essay. He continues: "[T]he dictatorship loathed (pretended to loathe) capitalism, yet, as in Russia, copied its methods, dictating names and slogans to the people with the same tenacity with which businesses impose their razor blades, cigarettes, or washing machines."[30] Notwithstanding, the use of propaganda was counterproductive in the end; Borges asserts that an excess of effigies of the political leader actually exacerbated the political struggle against him.

More striking for Borges was the manipulation of politics according to theatrical techniques such as "the rules of drama or melodrama."[31] An example of this manipulation was an event that took place on October 17, 1945, when "it was pretended that a colonel had been seized and abducted and that the people had rescued him. No one bothered to explain who had kidnapped him, or how his whereabouts were known."[32] Borges recasts a turning point in Argentinean history as theatrical farce. After Perón had to resign as vice president, he was incarcerated due to pressure exerted by the most conservative sectors of the army. Finally, a massive protest occupied May Square in front of the government building to demand his freedom. Four months after this event, Perón won the presidential election.[33]

Borges argues that it is useless to give a list of examples. He only seeks to denounce the ambiguity of the fictions put on stage by the abolished regime, "which can't be believed and were believed."[34] He also says that it is not possible to explain this contradiction by alluding to the public's lack of sophistication. He writes: "[T]he lies of a dictatorship are neither believed nor disbelieved; they pertain to an intermediate plane, and their purpose is to conceal or justify sordid or atrocious realities."[35] The Peronist regime, through the political manipulation of theatrical and propaganda techniques, induced the people to suspend their disbelief. It favored the collapse of the subject-fundament (*homo faber*) and, consequently, redefined the relationship between individual experience and action. It created a gap between individual cognitive capacity and mass behavior, and precipitated the fusion of individuals into a multitude. Thus, Borges believes that it is no longer possible to think about cognitively aware individual subjects in order to explain how the regime's *mise-en-scènes* were accepted.

In the short story "El simulacro" ["The Sham"], written in 1960 as a brief chronicle to mock one of the saddest events borne in the collective memory of Argentineans, Evita's funeral, Borges captures the reenactment that modern politics stages in this case to introject a theological element into the relationship between the sovereign and his/her constituency. The fact that he situates the story on a poor ranch in the province of Chaco—whereas the original ceremony in 1952 was a state funeral in the National Congress—instantly signals his grotesque portrayal of this historical event. He depicts Evita's dead body as a "blond doll" in a cardboard box and describes Perón "as tall, thin, Indian-like, with the inexpressive face of a mask or a dullard," who is mourning his wife while a procession of common people walk by to express their respect to the widower.[36] Borges's answer to the Peronist liturgy that sacralises the figure of Evita and deifies Perón's is to undermine any transcendental substance that attempts to legitimize either political figure. The story strikes at the pillars of the Platonic dualism that supports a third transcendental imaginary element that undergirds political representation.

By depicting Evita's dead body as a doll, Borges calls attention to the inert body of the political figure empty of the ideas that the people believed she represented. He also ambushes Platonic idealism through his depiction of the relationship between Perón and his constituency. The narrator says that the people address Perón to express their condolences "not for himself but rather for the person he represented or had already become."[37] They saw in him reflected the idea of what he represented, or perhaps it was no longer possible to distinguish between the original and its copy. Whether the former or the latter, it is interesting to identify in the text the third imaginary element that functions to capture the people and subordinate them under the political figure who reflected the Platonic ideal that they believe he represented. The funeral farce recast as a caricature aims to distort the relationship between the real political actors and the idea that they represent for the people. Nevertheless, Borges does not propose the imposition of a new transcendental order capable of investing ethical values in the relationship between the sovereign and his/her constituency. He only exposes this relationship from a refractory angle that allows us to perceive what is impossible for the political: the people in the story cannot break the enchantment—the Platonic spell—that has subjugated them.

In "La fiesta del monstruo" ["The Monster's Feast"] of 1947, which Borges co-wrote with Adolfo Bioy Casáres under the pseudonym H. Bustos Domecq, the opacity that characterizes Perón's constituency in Borges's texts becomes a virulent drive that animates a group of Peronist militants.[38] The narrator, who is a member of this group, recounts to his girlfriend the group's journey from working-class neighborhoods in the south of the Buenos Aires Province to a political rally at May Square. He is proud of his fusion with the group, which will then join with the multitude at the square, and of the murder that they committed with absolute impunity during their trip when a Jew they encountered refused to reverence the effigy of the "monster." The figure of the monster in the story alludes to Perón. Nevertheless, in this text the multitude converges with and within the "monster," becoming an irrational beast. "La fiesta del monstruo" provides a corrosive critique not only of the political, but also of the social movement that was represented by the Peronist regime.

Borges's literary way of thinking derives from a pallet that provides him with intense colors to delineate the tonalities and nuances of the impolitical dimension, which dwells in the texts that I have analyzed here. He radicalizes politics to reflect on the obscure consequences behind political representation. He points out how the political configures a particular relationship with *tekhnè* to manipulate the masses and the resulting irreversible collapse of the *homo faber*. He also clearly expresses an impolitical anti-theological attitude toward the sacralization of the political and undermines the Platonic idealism that invests political representation. Furthermore, he believes that a thanatic drive propels the multitude, the social actors who converge with and within the political.

Ground Zero: The Social and the Political

For both Borges and Esposito there is an emptiness that dwells within the political, a ground zero, an absolute negativity that defines contemporary politics and all forms of political representation. Even democracy, from their impolitical perspective, is never able to guarantee freedom; it only tends to forge a political relationship based on servility. In this sense, Esposito writes: "There is no real alternative to power" (*CIM*, 20). Borges, on the other hand, in the prologue to *La moneda de oro* [*The Golden Coin*] says: "I do not believe in democracy, that strange abuse of statistics."[39] Borges's and Esposito's impolitical attitude forecloses any possibility for the political to serve as an alternative to power and domination, and reveals the indissolubility of power relationships. This is the tragic and inexorable development of modern politics, which can only be based on oppression. Modern struggles for emancipation inevitably lead to the reenactment of previous forms of subjugation or, through technological innovations, create new systems with which to discipline and control the populace.

In spite of his various political positions, Borges is always consistent in his strong retreat from any collective political project. He ignores political subjects (groups, classes, masses, societies, and nations) as valid political bodies in order to give shape to an alternative politics. He rejects collective political movements that aim to appropriate the power of the State as a means to achieve emancipation; instead, he favors the power and the autonomy of the individual. Thus, he rejects political representation because it entails a violent reduction of the individual's private and internal sphere.[40] He disavows the cult of the State and, as he champions a defense of the individual, he assumes the impossibility of any actual political alternative.

In several of his short stories, Borges expands his defense of the individual into a collective intervention into the functioning of society at large. However, he manages to deny the State and its institutions as the crystallization of this expansion or as apparatuses through which the population should be governed. Thus, secret societies and conspiracies are crucial tropes in his literary work. In his texts, these clandestine organizations gather together a chosen minority of equals who share the secret of an order, whether divine or conceived by men, with which they surreptitiously infiltrate the world without violently fusing the singularities of each individual under a common ideal.[41] Bosteels claims that these closed groups operate in Borges's texts as an alternative mode of political intervention between the individual and the State, which represent the false extremes of the liberal-contractual theory of subjecthood and sovereignty.[42]

Esposito not only shares with Borges a similar repudiation of any devotion to the State, but also of any idolatry of the social, which also subjugates the person under the dominion of the political. In *Categories of the Impolitical*, Esposito begins, through his reading of Simone Weil's essays, to ruminate upon the concept of the impersonal as one key that can open a breach between the personal and the political.[43] Timothy Campbell points out that for Esposito the person (as well as the personal) is already embedded in the political, that "it is the principal mode by which the political takes form."[44] According to Esposito, Weil's critique of political parties, majority-driven democracy, and representation stems from her rejection of the myth of historical development—a Platonic dualism that assumes that matter is perfectible to an ideal form or that an ideal form has the power to shape matter. She believes that this conflation of the infinite realm of thought with the finite actual world is what defines idolatry. More precisely, she defines idolatry as the belief that there is a divinity that ubiquitously exerts its dominion.[45]

Weil further radicalizes this argument to define the social as idolatry. The social considered as the present and future of a community is already enmeshed in an understanding of itself in terms of the good delivered through an idolatry machine, which violently includes every relation within itself as belonging to it.[46] Campbell clarifies: "Esposito's close reading of Weil attempts to show [. . .] that the absolutization of the social as the good blocks other kinds of relations from coming into view, ones with other living phenomena; when they do emerge [. . .] the social reinscribes them in a relation between persons (and hence are once again social)."[47] This compression of relationality is the consequence of the idolatry of the social, which cannot escape from its devotion to the State that, since modernity, has emerged as the embodiment of power capable of regulating social relations.

The cornerstone of the state's juridical structure is the person who is invested with sacred rights by the idolatry machine. However, Esposito points out, "law is always of a part, of a party [but] never of all. All does not require law" (*CIM*, 238). It divides the whole into parts and pretends to equally guarantee the rights of each individual while, in fact, it violently enables individuals who have more power to have more rights. Hence, the triad of person-law-violence is synthesized under the myth of historical development, which Esposito considers to be analogous to the myth of Providence "as the personal will of God" (*CIM*, 240). He writes: "Providence as the historical form of divine pedagogy is nothing other than the making collective the metaphysics of the person; its expansion into becoming" (*CIM*, 240). The becoming of history is considered here as the object of Providence, which not only limits historical development but also governs all of its facets by providing norms and granting exceptions, shaping a form for historical becoming that is codified into law and order.

The concept of the impersonal inserts a wedge in the becoming of history as the object of Providence. In Esposito's early reading of Weil, he points out a line of fracture that moves toward an alternative relationality that is not confined by the social and resists being rectified by the idolatry machine. Against the theological-political semantic of the person, which captures the personal under the sphere of the political, the impersonal represents a positive pole. Nevertheless, this displacement is still an impolitical movement because it does not abandon its positive inoperativeness to provide an alternative to politics. Although it moves toward the outside of the social and the political, the impersonal drift continues to nourish a differential tension between inside and outside, without crossing the external boundaries of the political.

Esposito will then start to explore the idea that all political categories—sovereignty, representation, community, person—as defined by Hobbes and subsequent authors both authoritarian or liberal are merely a linguistic and conceptual mode with which to translate in philosophical-political and juridical terms the biopolitical *dispositif* that dominates the biological life of the populace in order to immunize it.[48] He draws these lines of analysis in *Categories of the Impolitical*, but he will later develop them, articulating a different genealogy of authors, following his books *Communitas* and *Immunitas*. Esposito says that he had perceived deconstruction as reaching a point of extenuation, whereas his philosophy was moving toward a horizon opened first by Nietzsche, and later by Foucault and Deleuze, "a horizon that refers to the political, economic, and juridical relevance of biological life as such."[49] First in *Bios* and then in *Third Person*, he asserts an affirmative biopolitics, and through a further elaboration of the concept of the impersonal, an alternative process of individuation that can unlock the current *dispositif* that captures individuals—their flesh and their biological life.

Notes

1. Esposito's *Categorie dell'impollitico* (Bologna: Il Mulino, 1988) was re-edited with a new preface in 1999 (*Categorie dell'impolitico* (Bologna: Il Mulino, 1999)). This preface was translated into English and published in 2009 (Roberto

Esposito, "Preface to *Categories of the Impolitical*," trans. Connal Parsley, *diacritics* 39, no. 2 (2009): 99–115). Esposito's book was translated first into French (*Catégories de l'impolitique* (Paris: Seuil, 2005)), and then into Spanish (*Categorías de lo impolítico* (Buenos Aires: Katz Editores, 2008)). Finally, Fordham University Press published *Categories of the Impolitical* (2015) translated by Connal Parsley.

2. Massimo Cacciari, "L'impolitico nietzscheano," *Il libro del filosofo* (Rome: Savelli, 1978). For an English translation see Massimo Cacciari, "Nietzsche and the Unpolitical," in *The Unpolitical: On the Radical Critique of Political Reason*, ed. Alessandro Carrera (New York: Fordham University Press, 2009), 92–103.

3. The impolitical perspective was elaborated in Italy during much the same time that Jean-Luc Nancy and Philipe Lacoue-Labarthe in France developed the idea of the "withdrawal of the political," which inspired them in their edition of the following books: Philippe Lacoue-Labarthe and Jean-Luc Nancy, eds., *Rejouer le politique* (Paris: Editions Galilée, 1981); and Philippe Lacoue-Labarthe and Jean-Luc Nancy, *Le retrait du politique* (Paris: Editions Galilée, 1983). Esposito recognizes the strong resonances between the impolitical and the work of these writers, although he asserts they were elaborated independently. He also points out that Jacques Derrida's work on deconstruction was another source that nourished the conception of the impolitical perspective (see Roberto Esposito and Jean-Luc Nancy, "Dialogue on the Philosophy to Come," trans. Timothy Campbell, *Minnesota Review* 75, no. 1 (2010): 75).

4. See Roberto Esposito, "L'impolitico," *Encyclopedia multimediale delle scienze filosofiche*, http://www.emsf.rai.it/aforismi/aforismi.asp?d=40 (accessed May 2014).

5. Alberto Moreiras argues that Borges's apathetic attitude regarding politics is an infrapolitical (*lo infrapolítico* in Spanish) stance "against the biopolitical rapture of politics" (Alberto Moreiras, "The Villain at the Center: Infrapolitical Borges," *Clcweb: Comparative Literature and Culture* 4, no. 2 (2002): 12). Moreiras first introduces the notion of the infrapolitical, which has a strong resonance with Esposito's earlier elaboration of the impolitical, to propose a theoretical articulation of Borges's poetics, before he expanded this concept in his book *Línea de sombra. El no sujeto de lo político* (Chile: Palinodia, 2006). Raúl Antelo also explores an impolitical dimension in Borges's texts, although he draws a different genealogy of impolitical authors than Esposito (see Raúl Antelo, "Borges y la impolítica," in Juan P. Davobe, ed., *Jorge Luis Borges: Políticas de la literatura* (Pittsburgh: Instituto Internacional de Literatura Iberoamericana, Universidad de Pittsburgh, 2008), 251–70). Diego Tatián also adopts the concept of the impolitical, which seems to have a similar meaning in his writing that it has in Esposito's work, to reflect on Borges's individualism (see Diego Tatián, "La habitación del monstruo," in *La conjura de los justos: Borges y la ciudad de los hombres* (Ciudad Autónoma de Buenos Aires: Las Cuarenta, 2009), 94).

6. Esposito, "Preface to *Categories of the Impolitical*," 102.

7. Ibid., 104.

8. Esposito quoted in Bruno Bosteels, "Politics, Infrapolitics, and the Impolitical: Notes on the Thought of Roberto Esposito and Alberto Moreiras," *CR: The New Centennial Review* 10, no. 2 (2011): 221. The original source is Roberto Esposito, "Por un pensamiento de lo impolítico," in Paolo D'Arcais Flores et al., *Modernidad y política: Izquierda, individuo y democracia* (Caracas: Nueva Sociedad, 1995), 226.

9. Esposito, "Preface to *Categories of the Impolitical*," 104.

10. Thomas Mann's book *Betrachtungen eines Unpolitischen* (Berlin: S. Fischer, 1918) [translated as *Considerazioni di un impolitico* in Italian and *Reflections of a Nonpolitical Man* in English], and Carl Schmitt's book *The Concept of the Political* (1932) are two main reference points for Esposito's impolitical perspective. However, he departs from the intersection of these two books through his reading of Massimo Cacciari's essay "L'impolitico nietzscheano" ["Nietzsche and the Unpolitical"] (1978) to develop his own understanding of the impolitical. Inna Viriasova has already suggested that Esposito's disclaimer regarding Schmitt's work is fundamental for our understanding of his impolitical perspective (Inna Viriasova, *Life Beyond Politics: Toward the Notion of the Unpolitical* (London, ON: UWO Electronic Thesis and Dissertation Repository, 2013), 139–40; http://ir.lib.uwo.ca/cgi/viewcontent.cgi?article=2314&context=etd (accessed May 2014).

11. Bosteels, "Politics, Infrapolitics, and the Impolitical," 221. Also see Carl Schmitt, *The Concept of the Political* (Chicago: University of Chicago Press, 2007), 26.

12. A Schmittian political theology formula is: "Sovereign is he who decides on the exception" (Carl Schmitt, *Political Theology: Four Chapters on the Concept of Sovereignty* (Chicago: University of Chicago Press, 2005), 5).

13. Carl Schmitt, *Roman Catholicism and Political Form* (Westport, CT: Greenwood Press, 1996), 21.

14. See Romano Guardini, *The End of the Modern World: A Search for Orientation* (New York: Sheed & Ward, 1956).

15. Quotations from this source are my translations.

16. Esposito quoted in Bosteels, "Politics, Infrapolitics, and the Impolitical," 220. The original source is Roberto Esposito, "Por un pensamiento de lo impolítico," 228.

17. See Schmitt, *Political Theology*.

18. Esposito relies on Arendt's understanding of politics in terms of a plurality impossible to represent. Politics can only be represented when it is violently reduced to a single symbol, which, however, still cannot subsume politics' intrinsic plurality. Arendt writes: "[A]ction, the only activity that goes on directly between men without the intermediary of things or matter, corresponds to the human condition of plurality [. . .] this plurality is specifically the condition—not only the *conditio sine qua non*, but the *conditio per quam*—of all political life" (Hannah Arendt, *The Human Condition* (Chicago: University of Chicago Press,

1958), 7). Also see Roberto Esposito, *La Pluralità Irrappresentabile: il pensiero politico di Hannah Arendt* [*The Unrepresentable Plurality: the political thought of Hannah Arendt*] (Urbino: Quattro Venti, 1987).

19. Esposito, "Preface to *Categories of the Impolitical*," 110.

20. Esposito and Nancy, "Dialogue on the Philosophy to Come," 84.

21. Esposito, "Preface to *Categories of the Impolitical*," 110.

22. I will not be able to engage here with the bulk of critical work regarding Borges's relationship with politics and his view on the political due to limited space of this chapter. I would like only to mention here that my reading of his texts has been influenced, among several other works, by Ricardo Piglia, *Crítica y ficción* (Buenos Aires: Seix Barral, 2000); Bruno Bosteels, "Manual de conjuradores: Jorge Luis Borges o la colectividad imposible," in Juan P. Davobe, ed., *Jorge Luis Borges: Políticas de la literatura* (Pittsburgh: Instituto Internacional de Literatura Iberoamericana, Universidad de Pittsburgh, 2008), 251–70; Bruno Bosteels, "La ideologia borgeana," *Acontecimiento: Revista para pensar la política* (1997): 51–92; and Annick Louis, *Borges, face au fascism* (La Courneuve: Aux lieux d'être, 2006).

23. Jorge Luis Borges, "The Dread Redeemer Lazarus Morell," in *A Universal History of Infamy* (New York: E. P. Dutton, 1972), 24.

24. Ibid.

25. Ibid., 25.

26. "The Dread Redeemer Lazarus Morell," which was published in the book *A Universal History of Infamy* in 1935, should also be inscribed in the specific historical context of Argentina during the 1930s. The first military *coup d'etat* in Argentinean history, led by generals José Félix Uriburu and Agustín P. Justo in 1930 against President Hipólito Yrigoyen, initiated the period known as the "Infamous decade," which was characterized by electoral fraud, political persecution, and corruption.

27. Jorge Luis Borges, "The Modesty of History," in *Other Inquisitions* (Austin: University of Texas Press, 2000), 167.

28. Jorge Luis Borges, "L'illusión comique," in *Selected Non-fictions* (New York: Viking, 1999), 409.

29. Ibid.

30. Ibid.

31. Ibid.

32. Ibid.

33. Borges points out that another example of the Peronist regime's theatrical staging took place on August 31, 1951, when the "dictator" (Perón) pretended to resign from the presidency (Borges, "L'illusión comique," 410). However, he did not present his resignation to the Congress, as would have been the proper procedure, but rather to union leaders. Consequently, Borges writes: "Crowds listlessly gathered in the Plaza de Mayo [. . .] Just before night fell, the dictator came out on a balcony of the Pink House [government building]. He was, as

expected, acclaimed, but he forgot to renounce his renunciation [. . .] Yet nothing happened that night: everyone (except, perhaps, the speaker) knew or sensed that it was all a fiction" (Ibid., 410). Borges claims that the purpose of this maneuver was to oblige the people to beg him to withdraw his resignation.

34. Borges, "L'illusión comique," 410.

35. Ibid.

36. Jorge Luis Borges, "The Sham," in *Dreamtigers* (Austin: University of Texas Press, 1964), 31.

37. Ibid.

38. Jorge Luis Borges and Casares A. Bioy, "The Monster's Feast," in *Chronicles of Bustos Domecq* (New York: Dutton, 1976).

39. Jorge Luis Borges, *Obras Completas* 3.121–22. After the horrendous crimes against humanity perpetrated by the military junta were publicly disclosed and democracy was reinstated in Argentina in 1983, Borges admitted being mistaken about his appreciation of democracy, an assertion that he had written on July 1976, only a few months after the *coup d'etat* that installed one of the bloodiest dictatorship in the history of South America (see María Esther Vázquez, *Borges, sus días y su tiempo* (Barcelona: J. Vergara; Editor: Grupo Zeta, 1999), 229–42).

40. Bosteels points out that Borges inherited from his father and his friend Macedonio Fernández, his political ideology: an anarchist and libertarian individualism, which essentially relies on Herbert Spencer's *The Man versus the State* (1884) (see Bosteels, "Manual de conjuradores," 252–53). For a further analysis of Borges' political variant positions, see Bosteels, "La ideologia borgeana."

41. For instance, see Borges's short stories: "Tlön, Uqbar, Orbis Tertius," "Theme of the Traitor and the Hero," "The Sect of the Phoenix," and "The Lamed Wufniks," or his poem "The Conspirers."

42. See Bosteels, "Manual de conjuradores."

43. Esposito focuses his analysis on Weil's texts on politics during the period before and after her essay "Human personality," which she wrote in 1943 (Simone Weil, "Human Personality," in *Selected Essays: 1934–1943* (Oxford: Oxford University Press, 1962).

44. Timothy Campbell, " 'Foucault was not a person': Idolatry and the Impersonal in Roberto Esposito's *Third Person,*" *CR: the New Centennial Review* 10, no. 2 (2010): 138.

45. See Simone Weil, *Attente de Dieu: Lettres et Reflexions* (Paris: La Colombe, 1950).

46. See Simone Weil, *La pesanteur et la grâce* (Paris: Librairie Plon, 1947), 65.

47. Campbell, "Foucault was not a person," 140.

48. Michel Foucault analyzes how in modern societies the power to decide matters of life and death exerted by the sovereign through the state apparatus assumes new forms after the nineteenth century. It becomes a disciplinary power and a biopower or biopolitics. Foucault writes: "[. . .] the existence in question is

no longer the juridical existence of sovereignty; at stake is the biological existence of a population" (Michel Foucault, "Right of Death and Power over Life," in *The History of Sexuality* (New York: Pantheon Books, 1978), 137). The biological life of living beings entered into the sphere of political techniques that institutionalize sexuality and administer and control the biological life of the population (also see Michel Foucault, *Security, Territory, Population: Lectures at the Collège de France, 1977–78* (Basingstoke, UK: Palgrave Macmillan, 2007); and Michel Foucault, *The Birth of Biopolitics: Lectures at the Collège De France, 1978–79* (Basingstoke, UK: Palgrave Macmillan, 2008). These texts, seminars, and lectures are the main groundwork upon which contemporary Italian philosophers developed their own understanding of the concept of biopolitics. For a clarification of the concept *dispositif*, see Gilles Deleuze's, "What Is a Dispositif?" in *Michel Foucault, Philosopher: Essays Translated from the French and German*, trans. and ed. by Timothy J. Armstrong (New York: Routledge, 1992); and Giorgio Agamben, *What Is an Apparatus?: And Other Essays* (Stanford, CA: Stanford University Press, 2009). I believe that Esposito's understanding of this concept relies on the former rather than on the latter.

49. See Timothy Campbell and Federico Luisetti, "On Contemporary French and Italian Political Philosophy: an Interview with Roberto Esposito," *Minnesota Review* 75, no. 1 (2010): 112.

10

Internalities of International Relations and the Politics of Externalities

Affirming the Impossibility of IR with Roberto Esposito

Mark F. N. Franke

International Relations (IR) theorists form an unusual movement in the discipline of Political Science. Outside of formal approaches, they tend to invite highly critical interventions into their own midst and commonly seek to apply or establish the pertinence of challenging insights they may draw from newly available philosophical work to the study of international politics. As a field, IR is regularly inviting of intellectual innovations. And, with the recent translations and expansive distributions of Roberto Esposito's writings, IR theory is now in the position to take in and attempt to both cultivate and apply what they have to offer. I reflect here on the stakes of such a moment.

Ostensibly, there should be much that brings Esposito and IR theorists together, both positively and negatively. Core topics taken up in his writings fuel standard debates in the field, and his claims in these regards are provocative if not menacing. Also, as one of the most influential thinkers on contemporary debates over biopolitics, there is a built-in attraction for Esposito's work amongst current investigations conducted by some who would name themselves critical IR scholars. Yet, the proposal that

I am making here is that the intellectual challenges offered by Esposito are not recognized if they gain reception only as either potentially for or against IR. They are neither. Rather, his writings allow for a thinking with both the theoretical traditions of and threats to IR that permits neither support nor rejection but only affirmation of the impossibility of IR as a theoretical or practical objective. Thus, I contend that Esposito's writings should not gain support or attention from within IR theory so much as they should be understood as challenges to the value of thinking politics in the terms of IR.

Esposito's work most immediately finds relation to IR in that it recalls the very political problems that provoke the historical theoretical moves around which this disciplinary project is built. In this respect, Esposito's readings of Thomas Hobbes's texts and theoretical functions are crucial. It is the same with his readings of Immanuel Kant's writings, insofar as Kant proposes solutions to the limits of Hobbes's strategies. However, the connection to IR in Esposito's work is not to be found in any evident refinements that he makes to how we may or must re-read Hobbes and Kant, allowing for a corrected IR, a new IR, or an anti-IR. Instead, drawing strong attention to what is normally forgotten by IR scholars about the theoretical inertia to be found in both writers' works, as the foundations of IR theory generally, Esposito gives fresh highlight to how the moves on which IR is based are themselves limiting of our engagements with politics. His writings give the possibility of a critical distance to thinking IR, a position of critique that is denied and avoided within IR itself.

The most prominent step that Esposito takes on this register is to trace the spatializing moves in terms of which the theoretical foundations of IR are organized. In reading the logic of state sovereignty that emerges through the writings of Hobbes and Kant, he brings relief to the contradiction of a tradition of thought that, at once, requires the interiorization of politics to forms of territory that can neither tolerate nor withstand the existence of an outside that remains external to the spaces of politics. Thus, Esposito's relation to IR is not to be established in terms of how we might situate him in the "Great Debates" of the field or any other formulation of the stakes claimed as proper to IR scholarship itself. His relation to the field is better understood as not pertaining to IR in any strict sense. Rather, in relation to IR, Esposito is a critical theorist of political theory who finds interesting problems in the core commitments to which IR subscribes that he understands can be productive of new political possibilities altogether. He offers a critique that avails itself of political effect relevant to the discipline, wherein IR is found and rendered less possible, even in a critically radicalized form, because his acts of critique provoke recognition of greater possibility in terms of thinking and doing politics.

Esposito's Relation with IR: Readings of Thomas Hobbes

The political imaginary that fosters IR is populated by a collection of communities or proto-communities, of varying degrees of strength and character, that confront each other in a singular and unified terrestrial world absent of any given conditions under which common order or governance is possible. IR starts from the assumption of community or its inevitability, as a positive political fact, from which to pose and analyze broader problems over how it is that communities may function politically, as distinct agents, in relation to each other. To this end, theoretically, IR relies very heavily on stock characterizations of sovereign community as they may be drawn from Hobbes's *Leviathan*, deriving a disposition by analogy from what he actually argues with respect to human beings who attempt to form community. By contrast, Esposito is well known for his argument that we cannot legitimately receive community as a positive object for political theory of any sort. Rather than given, he understands community to function in affirmation of a negativity, at the level of ontology, connoting a site of dynamic political struggle instead of a singularized agent of politics. However, this fact does not put Esposito so much at a distance from the theoretical core of IR, as traced to Hobbes, as it establishes commonality between the two. Rather, Esposito is far more perceptive of and in agreement with what Hobbes argues about community than is normally understood or given attention in IR, showing IR theory to be out of step with its own tradition.

IR theory seeks legitimacy in the ideas of several and highly varied political philosophers from over the past half-millennium, reducing their ideas to stock notions of historicism, realism, idealism, and liberalism and generating four key debates in these terms. Ultimately, the crux of each debate is some manner of struggle with apparent ambiguities in the lessons derived from the writings of Kant, but they are sustained by the core comfort found in the revolutionary modern proposals of Hobbes's *Leviathan*. This is not because Hobbes provides a steady, reliable, or persuasive foundational map to international politics in which all sides of the debate can find a home or footing. On the contrary, he provides no international theory of any sort. Hobbes only provokes and cultivates the problem of international politics, implying at the root of his writings that there are no given limits in which to think politics in the world beyond the communities that we might build, discursively, with one another.[1] He negates the greater world as a site of community and, thus, renders it an inescapable site of politics for communities, where communities must confront and act in relation to one with political arts alone. For Hobbes, it is only within the secured autonomy of community that political agency and life is realized.[2] And it is in this fundamental step that his thinking appears negated by that of Esposito.

Esposito explains that the very idea of community itself offers no reference to a positive unity of persons who, together, form a singular political subject in the world (*CM*, 1). His point is that community does not and cannot express a core property that its members share with one another in that sense. Rather, he argues that community is accomplished in each possible member subtracting what is proper to her- or himself and giving over to dynamic society with others that which would be self-identifying (*CM*, 3–11). Thus, Esposito determines community more in terms of a void, where there exists neither individual subject nor communal subject per se. He sees no public property in community, either, as a compensation for the loss of individual personal property.[3] Rather, it is a site at which persons have exposed and given over in common any force of self-determination for the purpose of being political with one another, as something they owe one another in community.

With apparent similarity, Hobbes understands sovereign community to be formed when individuals give over to one another their individual liberty in common, as a way of forging a singular political body for their mutual security and liberty, a common body that can reflect the personality of each member and affirm mutual political authority.[4] He sees community as a human possibility but not a true necessity of being human.[5] Thus, Hobbes and Esposito share in the observation that community is something that is not a given feature of the world but, rather, that it must be built. It is not a component that can be assumed as an established variable from which one might then build something like IR theory. Whatever one might want to develop as a theory of international politics, from either analysis, could not depart from the politics of community. In either case, community itself is theorized in such a highly contingent fashion that a purely conceived IR would be untenable. In either case, politics is not something resolved at the level of community. At best, community offers a condition of possibility for international politics, but a condition that is itself still political in its dynamics. And, despite the efforts of generations of IR scholars to do so with respect to the ideas of Hobbes, there is no reasonable manner by which an international politics could be thought beyond community. However, whereas devotees of Hobbes might hope that a set of communities could develop, historically, across the globe with such separation and stability as to make possible what IR theorists will to think, as an international political condition distinct from a stable register of community orders, Esposito offers no such hope at all.

Key to the difference that Esposito is expressing in his analysis is the fact that this understanding denies to community any form of enclosure. It has no bounded shell, from which an outside world or that which is not-community truly could be externalized and rendered into the realm of an international politics that the field of IR typically thinks it can derive from Hobbes's vision of a secured community. For Esposito, there is nothing to contain (*CM*, 138). Community is improper, offering an "inside" that is against the possibility of the proper.[6] Community itself is a void, made when the participants expose themselves to society ordered around nothing in particular. However, on this very point, he is not attempting to contradict Hobbes so much as he is seeking emphasis on a point that is fundamental to Hobbes's own analysis. Esposito draws his readers back to the point that the rationale for the artifice with which Hobbes understands sovereign community to be built is only the recognition that we may be killed (*CM*, 21–29). The core of sovereign community is nihilism, acceptance that we may come together because of the ever-present potential of our mutual negation. Thus, the improper is really never something that can be abandoned as such, and the supposed property of community is an illusory construction.[7] For Hobbes, it is in fact the willingness of individuals to give over in common this potential of mutual negation from which the sovereignty of community arises, in the form of a sovereign who holds the same sword over each member's neck equally.

As a site and practice of fundamental and common negation, then, Esposito's point is that community, in itself, cannot offer the world an inside against an outside. On the contrary, community as such can only be external, in that it does indeed expose all members to an extraordinary situation of openness. Rather than providing political solution to a common problem of conflict, community opens individuals to the problems of possibility and uncertainty that defer politics as a problem of conflict.

It does not offer place but, instead, reveals one to the dynamics of global spaces.[8] There is a doubling and inverting sacrifice, for Esposito, in the formation of community. Those who are not members of the community are sacrificed to the outside of it, but the internal political subjectivity of each individual member of community is sacrificed to an outside as well (*CM*, 33–34). The relations that members of the community have with one another are formed insofar as the individualities of the members themselves are moved to the outside (*CM*, 139). For Esposito, community provides to the world no political particular to be gathered in a common universe of international politics capable of sustaining anything like IR theory. Community externalizes the particular, rendering itself open to politics that cannot be categorized in terms of the world, international, or global, but, rather, only an outside to which community is immediately vulnerable.

Esposito's Displacement of IR as a Concern: Readings of Immanuel Kant

Exploring the other modern foundational theory of community, as these things are taken up in IR, Esposito traces the implications of Kant's arguments over the moral necessity of community in the state. While Kant accepts the practicality of the fear and nihilism that drives persons together to establish community, as articulated by Hobbes, he goes much further to outline also how the formation of territorially sovereign state community is both rationally and morally necessary.[9] Kant reads human history as an ongoing effort to establish state community and to globalize its form exhaustively.[10] Thus, unlike Hobbes, he does indeed leave room for a theory of international politics, supporting an idea of political community that ultimately must fall within a global order of other such communities. However, as Esposito emphasizes, Kant's global idea of international politics is not something to posit as existing as such. It remains only an ideal to guide one ethically within the context of international society.[11] And as a guide, Esposito contends, the value of community for Kant is necessarily empty and does not position anyone in relation to an international realm of politics typically idealized in IR (*CM*, 71). Rather, it propels one to actions of state internalization that serve to undermine the international as site of political possibility. Thus, for Esposito, insofar as we think and or attempt to practice community in the form of state sovereignty, rather than bringing the international to bear on us and give reason for IR theory, the question of IR is rendered irrelevant.

Kant introduces a highly positive understanding of what life under modern state sovereignty could become.[12] He idealizes the state as a site in which persons could enjoy a freedom of citizenship and work with and against one another in the imagining and building of community he believes could truly dignify citizens as free and equal human beings.[13] Yet, as Esposito argues, the public processes to which Kant refers in such moments of his writing do not in themselves describe a theory of community for Kant (*CM*, 62–71). Rather, they are morally grounded activities that are guided by an abstract ideal of community that could have no practical reality in itself. Esposito's claim is that the location of community for Kant is in no public realm at all but exists only in a form of law. His point is that Kant does not give us a reading of community in which lawful life can be produced but, rather, a reading of law in which community is made thinkable and which may be formed in application of the law (*CM*, 64–65). This is not to say that law gives us community. Instead, it is to take the position that the law may regulate social activities in such ways that community becomes theoretically realizable. However, as Esposito contends, for Kant community remains a matter of theory contingent on the law and never a possible practical reality that could be recognized and enjoyed as such (*CM*, 71).

The way in which Esposito characterizes Kant's articulation of community is as an impossibility. He states that, for Kant, it is not so much that community is impossible as it is that this impossibility is itself community (*CM*, 76–77). Thus, while for Hobbes persons are united in community by nothing in negation, for Kant they are united by what Esposito refers to as a "nothing-in-common" that stands in an oddly positive fashion for each person even prior to the social contract of the state (*CM*, 77). And, for Kant, it is this nothing that in reference to which the social dynamics of the sovereign state must be organized. The society of this state ideally is ruled by law-making that is premised upon and asserts a principle of community that cannot be achieved but must always serve as the guide for all persons in their relations with one another.

Concluding that the only thing that any one of us, as rational beings, can know certainly and pre-discursively is that we are free in our reason,[14] Kant surmises that all such beings not only ought to be free but that they have a duty to establish the conditions of such freedom, for themselves and one another. Yet, Kant argues further that such freedom is not something to which any of us can aim in an obviously positive and determinate way. Thus, he maintains that we must each approach freedom ethically, from how one ought to will one's actions for the possible attainment of individual freedom.[15] Framing this observation in terms of his categorical imperative, Kant sees it as necessary that the rational being seeking fulfillment of duty to his own freedom must understand that such freedom would be limited by the freedom of other such beings.[16] Thus, consideration of our social contexts and questions of social order are of close concern to him, but not as particular and unique goals. Rather, for Kant, the ordering of society with the limits and around the institutions of a state is primarily a formal process directed by adherence to a universalized and abstract law. The state is a place from which knowledge of a fully free community of rational beings could be attempted. It is not a site for such community itself.

Kant is interested in how the state makes possible the institution of social order on the basis of law but also the liberalization of laws that would permit the freedom of citizens of the state to explore with one another, on the basis of critical interrogation, how it is that social and legal order could be reformed to the point where the categorical imperative could indeed be said to rule.[17] However, the development of social order on the basis of such right reason[18] in the institutions and public discourse of the state is insufficient for Kant, given that the categorical imperative must accommodate the freedoms of all possible rational beings, and it must be based on the register of critique rendered possible in all such states. The law that can give one view to community must be cosmopolitan in basis. However, recognizing that it is unlikely that such a complexly common social legal order could and would be willed by all rational beings equally on a planetary scale, Kant is under no illusion that the species will ever be under the legal conditions within which the community of freedom could finally be thought in any true sense.[19] Kant cares only that we orient ourselves toward an idea of community, regardless of its impossibility, as a way of dignifying the knowledge that we ought to be free.

For Esposito, then, the orientation that Kant demands of us is neither toward the inside nor the outside, nor toward relations between the two. It is an orientation directed toward elimination of the distinction between particular societies in state formations and the realm between their borders. Kant's reading of community looks toward the legal rendering of the international internal to the order of each state, where the international is domesticated altogether. This is what Esposito argues about the law generally, that it serves to eliminate the outside (*IM*, 30–31). And this is true to the core of Kant's reasoning, where he requires the globalization of state formations that, in turn, seek internalization of the threats that they pose to one another, internationally, through the universalization of legal forms. Thus, as a ground for IR, Kant's objectives offer neither community nor the international. These objectives idealize community's impossibility and set about political orders that seek the end of politics.

Negating IR: A Theory of Immunity

What Esposito traces as the internalization of the outside by law, in the state, as is typified in the writings of Kant, is connected directly to Esposito's broader theory of immunity. He takes the position that the positive and contained idea of community, on which IR relies, is thinkable and presentable only insofar as the supposed inside takes in the threats of the supposed outside (*IM*, 16). There are two actions at stake in this regard. First, by performing the action of inclusion or taking-in the state literally describes a division between inside and outside that is not already given. It is a performance of the limits of its own description. Second, by incorporating parts designate of the outside that might otherwise serve as threats to the community projected by the state, the state can perform acts of familiarization with its projected outside and propose itself as a set of mechanisms capable of managing that which threatens it. The outside is rendered as objects for the knowledge and manipulation of the subject-state. Just as law gives sovereignty to the state and a theory of community-oriented society, the opening of law to the laws of other states, in the practices of international law, simultaneously permits or takes-in the influence of other jurisdiction, as external rules to be considered inside, and renders the laws of other states subject to domestic law-making practices. The implication is that what is encountered, negotiated, and described as international politics through law is the internalization of relations, rather than the participation in International Relations per se. From this perspective, then, the prospect of IR is a domestic order for every state and cannot be recognized as that which scholars of IR would claim it to be.

The state's move to the international via law, for Esposito, must be an activity of internalization of the foreign, as opposed to a confrontation with the foreign, precisely because, at the level of community, the state has only a lack or a nothing from which to build itself. And he surmises that the negation around which the state attempts to build its position in a world can become such building material only through its own negation. Just as in formal logic, where the negation of a negative statement is equivalent to the positive form of the latter, Esposito argues that a positive predication of something like state-based community may be enacted only insofar as its fundamental negation, its lack, takes on its own negation (*IM*, 8). For the nothing to appear as something, the nothing itself must enjoy negation, such that the nothing is given substantial definition. The commonality of the nothing in common must be inverted to a common nothingness, and the open with which the nothing of community is faced, from the "outside," must be rendered into something around which the predicated community may give itself singular form (*IM*, 21–28).

With the understanding that community is a form of life whose commonality is the fact that all members may be killed, that they are vulnerable to one another and the world, Esposito argues that the key feature that allows the modern state to predicate its own society as a form of closed community is through an inversion of the commonality of death to a common capacity to kill. Community is formed in an immunity system folded back on itself (*IM*, 51). In this regard, he takes up the puzzle left by Michel Foucault over how the modern politics of life so easily reverts to a politics of death.[20] His point is that the dynamic of modern sovereignty is to establish an enclosure of its own society and mechanisms around which it could bring death from the larger outside, to learn about these threats and to learn how to eliminate them. In this way, a particularized form of life may be enacted against others, where outside threats themselves are particularized as objects for the killing or neutralizing apparatuses of state society. Yet, such focused acts of immunization are possible only with immunization from within as well.

To invite and achieve a negation of the negation of community, Esposito posits that the very negation with which persons in the society meet one another also must be negated. They must be able to separate from their common condition of a lack to produce a kind of subjectivity capable of negation, wherein the common negation is individuated.[21] He understands that a form of state-oriented individual subjectivity is required, wherein each person of that society is encouraged to negate the threat that they pose to one another. Immunization involves violent revolt against oneself,[22] through forms of selective counterinsurgency that cut across the lines of domestic and foreign.[23] "The 'war of all against no thing' replaces the [Hobbes's] 'war of all against all.'"[24]

The ultimate consequences that Esposito is tracing in this dynamic is that the sovereign state can accomplish its predication of a positive form of community only insofar as the outside becomes inside and the very negative basis of community, where persons are open to each other, is itself negated. Alterity is not opposed but is neutralized as such.[25] Further to this point, he is indicating how the individuation of persons under the law of state sovereignty and its programs of immunization are inverting of subjects to one another, where the individual self is established only insofar as she or he takes on possible others as its own inner substance. In these ways, Esposito is not trying to describe the dynamics of modern state sovereignty as an opposition between state politics and the life that it seeks to protect and promote. Instead, he describes what he sees as the intersection of the two. This is particularly so under conditions of globalization, the increased communications between and intersections amongst humans, ideas, languages, and technologies.[26] Sovereignty does not require the contradiction of life by politics. Rather, the politics of sovereignty works with life but produces contradictions in life (*IM*, 138–39). They may be read together within the paradigm of immunity.[27] This is the core legacy of Hobbes.[28] And, in this regard, Esposito suggests that immunization is not a mere tool of modernity but, rather, is what brings about modernity as a possibility.[29] The modern state bears with it not simply the capacity of immunity but it is also a product of its conflicting dynamics. Thus, Esposito is preoccupied with a central complex of questions: "is there a point at which the dialectical circuit between the protection and negation of life can be interrupted, or at least problematized? Can life be preserved in some other form than that of its negative protection?" (*IM*, 6). And, in response, he is not anticipating the negation of immunization itself but, rather, the productive capacity of the deepening practices of immunization.

Esposito takes the position that the negative forces of immunization are themselves stimulating of what they negate: the common. The politics of immunization, for him, are not a battle between an original form and a foreign form of being. Instead, as Esposito sees in the example of what is typically depicted as the recent global war lead by a secular West against Islamic fundamentalism, this is a politics of transformation.[30] He is interested in how immunization brings about alterations (*IM*, 174). Specifically, then, rather than subdue the life evident in the fundamental core of community, its lack, Esposito offers a more radical position.[31] He argues that immunization emphasizes the commonality of life and its mutual intersection in each of us, in rendering an inside that must be outside in order to be inside, and in rendering an individual subject that must be other in order to be self. The immunitarian paradigm follows an expansive liberalism rather than exceptionalism.[32] In this way, he imagines that the force of lawful sovereign politics of the state must provoke life in ways that encourage us to ultimately stand in common with one another. If anything, this politics renders a commonality manifestly evident and unavoidable.

From the Apolitical Humanism of IR to the Politics of the Impersonal

Esposito's will to cease thinking of life as a function of politics and, rather, to understand how politics is better thought as within the form of life itself appears to have a point of direct contact and overlap with the most overt of disciplinary forms in contemporary IR (*BS*, 12). The claims to territorial communal sovereignty on which the now largely Kantian IR is ordered are idealized and justified largely in terms of states' capacities to order and serve the protection of internationally recognized rights and freedoms of citizens, as human beings, rather than the parochial interests of respective states themselves. Increasingly, IR grounds itself on the universalized premise of human life. Yet, Esposito is quick to point out that the kind of retreat to a basic human life now characteristic of IR offers no resting place for politics. On the contrary, for him, the move to both ground and regulate sovereign claims in terms of the rights supposedly proper to human life per se is fundamentally de-politicizing. And the stress of this condition of de-politicization itself must produce a condition of political life that disallows for any simple relation between universal and particular, as would be required by IR.

Under the system of the United Nations, states seek legitimacy in their politics by asserting their legal will and capacities to protect humanity in the form of what is taken as fundamental to the life of each individual human being. Yet, Esposito reminds us that the address of the human in law, especially within the context of rights law, does not pertain to the merely alive human being, in its specificity, but, rather, to the subject of law, the person. While he acknowledges that the notion of the person is intended to address and dignify human beings complexly, in their humanity, their lives, their political identities, bodies, and even souls, Esposito questions how this form of legal personalization can possibly achieve its aims. And, in recognizing the many gross failures of post-World War II universalist human rights law to provide adequate tools for the politics of protecting human life, he argues that the common disconnect between protective rights and the protection of life in the world is not a result of poor applications of the law but, rather, is symptomatic of the direct and expansive application of the ideology of the person inherent to it (*TP*, 4–5, 68–69).

Building an analysis of legal personality, as it is derived from Roman law, Esposito points to the fact that the human who is celebrated in the Universal Declaration of Human Rights, around which the whole UN system revolves, is a person whose primary characteristic is the possession and exercise of sovereignty over her or his own body, analogous to the relation between state and territory and citizens (*TP*, 10). The human person, in this regard, is presentable insofar as she or he can gain dominion over her or his living corporeal being as an object for his or her own subjectivity (*TP*, 11, 82). The human subject of rights is one who can rise above her or his own material being, as a form of animal being distant from her or his legal form (*TP*, 12). It represents the split between biology and rationality (*TP*, 83), where the former is rendered an object for the latter,[33] and the life of the person is protected against a freed animal body.[34] The human of human rights then can only be a generic form of being incapable of bringing in the individual living being (*TP*, 67). It is a site of being divided from its life (*TP*, 74).

Rather than considering and analyzing problems that states have had in protecting life, according to the law, then, Esposito is initially concerned to show how the international legal regime which gives states sovereign rights and duties to protect human beings on the basis of universalized human rights renders states themselves as the agents responsible for abstracting humans from their lived lives. Insofar as the sovereign establishes personality for the state itself, it provides for a standard of legal personality that can only be generic and abstract. Again, to return to Hobbes's analysis of sovereign community, the personality of that community is established in the negation of individuality toward the lack of anything in common. The standard of legal personality in that order is an empty position in which no one individual living being could persist as such. Thus, Esposito contends, the sovereign functions both to personalize its citizens, in transforming each into a person before the law who may claim rights with respect to human properties that overcome particularities of the body, and to de-personalize its citizens, in transforming the body of each subject of law into something insufficient for standing before the law (*TP*, 87). And the consequence of such sovereign action is to establish the person as a site for which humans might compete for the right to claim, where one might seek to demonstrate her or his humanity against another's lesser humanity or animality.

At a more profound level, though, Esposito refuses the idea that the human person, as a truly autonomous being, can be accomplished or found in either particular or universal form. He rejects the existence of a "human nature" as such.[35] What leads Esposito to this manner of thinking is a fundamental deconstruction he traces in the modern subject of rights, the person (*TP*, 104–09). This human being must be able to present itself before the law. It must be able to articulate its own "I" that is self-generated. Yet, it is impossible to declare anything like "I am here," in the first person, without also being able to posit the "you," of the second person, to whom the "I" addresses and presents itself. In declaring its own self-mastery, over its body and in its own rationality, the human subject of rights indicates its very inability to actually present itself, in that its presence must be affirmed in the second person, in the "you." Similarly, the second person can give presence to the first person only insofar as the second person also affirms a first-person position, in which she or he can state that "yes, I see that you are a human being and, on that basis, have a right to rights." Thus, the supposed autonomy of the subject of rights is caught within a reversal of origins to which neither "I" nor "you" is adequate and to which the dialectic between the two is also inadequate. The humanity and rights of any articulation of the first person as well as its affirmation in the second person are displaced elsewhere. And Esposito traces this subjectivity to a site that makes both possible and impossible at the same time, the third person.

For the first and the second person to be able to assess one another, in their reversing forms, they must have recall to the idea and category of personality that could hold for both and all forms of persons. In their interrelations they must have recourse to a reference for human subjectivity that holds for all but for no one in particular. This is found in the form of the third person, which is then impersonal. And, as such, the ground of any claim to human rights, in the forms of either being a subject of rights or acknowledging the autonomy in another, lacks the singularity, sovereignty, or individuality that is asserted in such human personality to begin with. As Esposito notes, whereas the rightful person asserts an interior being contained and disciplined by its self-sovereignty, the impersonal third person from which such an articulation is made possible is without interior (*TP*, 133–42). It is only externality. The ultimate source of the first person is the possible contact of every possible person with one another and their inherent openness to one another as living beings. The possibility of the human person is premised on the impossibility of such a person, in that such personality cannot be confined to itself and is always generated as possible in its life with others.

Thinking Politics; Not Thinking IR

Whereas Kant would have the thinking of politics, as a necessarily international politics, guided by the idea of the state and the goal of a planet-wide lawful community of states, Esposito contends that we cannot actually theorize politics from within such limits. For him, the kinds of practical and theoretical disciplinary confines provided by IR give us no view to politics.[36] Esposito does not give us a new perspective for IR but, rather, challenges us to accept and affirm the contradictions of state sovereignty and the self-sovereign human citizen that may make possible manners of being that do not fit into the logic of an international community of communities. While IR theory, in all its forms, posits relations between given or necessary communities and subjects, Esposito sees the value of accepting forms of life that are fundamentally relational and impersonal.[37] It is in this externalized exposure to one another and life overall, where the presuppositions of sovereign mechanisms are revoked,[38] that it becomes possible to open thinking to how we *may* be with one another, rather than how we *must* be with respect to one another. Esposito points to opportunities in which we may think about politics rather than pretend that IR amounts to anything in the least political.

Notes

1. Thomas Hobbes, *Leviathan*, revised student edition, ed. Richard Tuck (Cambridge: Cambridge University Press, 1996), 47–49.

2. Ibid., 88–89.

3. Greg Bird, "Roberto Esposito's Deontological Communal Contract," *Angelaki: Journal of the Theoretical Humanities* 18, no. 3 (2013): 38.

4. Hobbes, *Leviathan*, 91–100.

5. Ibid., 117–21.

6. Bird, "Roberto Esposito's Deontological Communal Contract," 34.

7. Ignass Devisch, "How (Not) to Properly Abandon the Improper?" *Angelaki: Journal of the Theoretical Humanities* 18, no. 3 (2013): 77.

8. Bird, "Roberto Esposito's Deontological Communal Contract," 44.

9. Immanuel Kant, *Anthropology, History, and Education*, ed. Günter Zöller and Robert B. Louden (Cambridge: Cambridge University Press, 2007), [8:116–18] 169–71; Immanuel Kant, *Practical Philosophy*, trans. and ed. Mary J. Gregor (Cambridge: Cambridge University Press, 1996), [6:334] 475, [6:343–48] 482–85, [8:347, 8:354–56, 8:376] 320, 325–27, 343.

10. Kant, *Practical Philosophy*, [8:343–86] 317–51; Kant, *Anthropology, History, and Education*, [8:17–31] 108–20, [8:107–23] 163–75.

11. Kant, *Anthropology, History, and Education*, [8:18–22] 109–12.

12. Kant, *Practical Philosophy*, [8:35–42] 17–22, [6:311–42] 455–80.

13. Ibid., [6:311–42] 455–81.

14. Kant, *Practical Philosophy*, [8:361–62] 331–32; Immanuel Kant, *Critique of Pure Reason*, trans. Paul Guyer and Allen W. Wood (Cambridge: Cambridge University Press, 1998), [A114, A802/B830–A803/B831] 236, 675–76; Kant, *Anthropology, History, and Education*, [8:19, 7:329] 110, 424–25.

15. Kant, *Practical Philosophy*, [8:275–309] 279–309.

16. Ibid., [4:421] 73, [6:318–20, 6:331–37, 6:366, 6:381] 461–63, 473–77, 500–01, 513–14, [8:299] 298.

17. Ibid., [6:311–32] 455–506.

18. Ibid., [6:229–372] 386–506.

19. Kant, *Practical Philosophy*, [6:350] 487–88; Kant, *Anthropology, History, and Education*, [7:311] 427.

20. Roger Cooter and Claudia Stein, "Cracking Biopower," *History of the Human Sciences* 23, no. 2 (2010): 118–19.

21. Roberto Esposito and Timothy Campbell, "Interview," trans. Anna Paparcone, *diacritics* 36, no. 2 (2006): 50–51.

22. Massimo Donà, Loredana Copmarone, and Andrea Righi, "Immunity and Negation: On Possible Developments of the Theses Outlined in Roberto Esposito's Immunitas," *diacritics* 36, no. 2 (2006): 65.

23. Colleen Bell, "Hybrid Warfare and Its Metaphors," *Humanity: An International Journal of Human Rights, Humanitarianism, and Development* 3, no. 2 (2012): 225–47.

24. Brett Levinson, "Biopolitics in Balance: Esposito's Response to Foucault," *CR: The New Centennial Review* 10, no. 2 (2010): 240.

25. Donà et al., "Immunity and Negation," 62.

26. Esposito and Campbell, "Interview," 54–55.

27. Adalgiso Amendola, "The Law of the Living: Material for Hypothesizing the Biojuridical," *Law, Culture and the Humanities* 8, no. 1 (2012): 114.

28. Bird, "Roberto Esposito's Deontological Communal Contract," 37.

29. Esposito and Campbell, "Interview," 54.

30. Ibid., 53.

31. Cooter and Stein, "Cracking Biopower," 122.

32. Joshua Barkan, "Roberto Esposito's Political Biology and Corporate Forms of Life," *Law, Culture and the Humanities* 8, no. 1 (2012): 91–92.

33. Timothy Campbell, "'Enough of a Self': Esposito's Impersonal Biopolitics," *Law, Culture and the Humanities* 8, no. 1 (2012): 37.

34. Jaakko Ailio, "Liberal Thanatopolitics and the HIV/AIDS Pandemic," *Alternatives: Global, Local, Political* 38, no. 3 (2013): 260.

35. Timothy Campbell, Federico Luisetti, Roberto Esposito, "On Contemporary French and Italian Political Philosophy: An Interview with Roberto Esposito," *Minnesota Review* 75 (2010): 113.

36. Bruno Bosteels, "Politics, Infrapolitics, and the Impolitical: Notes on the Thought of Roberto Esposito and Alberto Moreiras," *CR: The New Centennial Review* 10, no. 2 (2010): 224.

37. Timothy Campbell, "'Foucault was not a person': Idolatry and the Impersonal in Roberto Esposito's Third Person," *CR: The New Centennial Review* 10, no. 2 (2010): 141–42.

38. Roberto Esposito, Jean-Luc Nancy, and Timothy Campbell, "Dialogue on the Philosophy to Come," *Minnesota Review* 75 (2010): 77.

11

Living Through Catastrophe

Warring Immunities, Dramatization and Counter-Actualization in Wajdi Mouawad's *Scorched*

Geoffrey Whitehall

> Who knows . . . Brothers are shooting brothers and fathers are shooting their fathers. A war. But what war? One day 500 000 refugees arrived from the other side of the border and said: "they have chased us off our land, let us live side by side." Some people from here said yes, some people from here said no, some people from here fled. Millions of destinies. And no one knows who is shooting whom or why. It is war.
>
> —Wajdi Mouawad

Wajdi Mouawad's 2003 play *Scorched* is about life, resistance, and living through catastrophic times (see also Denis Villeneuve's film adaptation *Incendies*).[1] It is first and foremost a catastrophe play. It is driven toward *the catastrophe*, the second to last scene that brings the confusion, intrigue, and drama to a resolution in the finale of the play. The catastrophe leads to the shock of revelation that, in the end, defines the resolution. Because of its resolution, *Scorched* has similarities with a tragedy. Both simultaneously explain everything and yet change nothing. The audience is left stupefied: such trauma, hurt, anger, panic, fear, disgust, and loss![2] However, *Scorched* is also a political play. Instead of letting catastrophe frame the resolution in terms of tragedy's contemplation, *Scorched* seeks lived transformation.

It asks: How can we live after trauma? How can we resist the drama of catastrophe so that new lives can emerge? Instead of just surviving life, *Scorched* demands that life thrive in and through catastrophe.

This chapter mobilizes the work of Roberto Esposito through a reading of *Scorched*. Esposito offers us a more useful way to depict the catastrophe of international conflict by shifting the geopolitical imaginary from clashing essential communities (like Nations) toward complex biopolitical notions of shared and entangled immunities (which I call warring immunities). Furthermore, I show how Esposito offers a way to reconceptualize the objective goals of our catastrophic times by moving away from discourses of security and survival toward the transformation of lives lived. The political objective shifts from deferring or preventing catastrophe toward acknowledging and transforming the catastrophe in which we already live. Under these conditions, the chapter rejects Esposito's reliance on passive biological processes to bring about transformation and instead pushes toward the active political transformation of life and love. By making life and love a political promise via dramatization and counter-actualization, the theatre of *Scorched* is shown to be a political guide in our catastrophic times.

Scorched begins with the reading of Nawal's *will and testament*. In her final years, Nawal has fallen silent and has not spoken to anyone—something catastrophic has overwhelmed her desire to continue fighting. She left written instructions, upon her death, to be buried in a hole, face down, naked, with no coffin or epitaph and only cold water to be thrown on her grave. What could justify such a silent burial? Amplifying the mystery, she leaves two letters to her twins—Janine and Simon. The letters do not explain Nawal's condition. The twins are not to open the letters. Janine's letter is to be delivered to their unknown father and Simon's letter is to be delivered to their unknown brother. Prior to receiving the letters, the twins did not know that they even had a living father or brother. In order to deliver the letters, as such, they must leave their home in Quebec for Nawal's unnamed birthplace. On the journey, they are to discover the tumultuous secrets of Nawal's life. This journey reveals not one catastrophe but a series of catastrophes. It reveals Nawal's incredible resilience but, and this is the premise I explore in this paper, the play cumulates in a transformational Event. The play is not about Nawal's struggle or her surviving great hardship; it is about the political audacity to transform the conditions of the past-future and future-past so that new lived horizons can thrive.

Although Mouawad allows the audience to make their own connections beyond the idealized geography and history of the play, it is well known to be set in the Lebanese Civil War in 1975–1990 that left over 100,000 dead and 1,000,000 displaced. Also known as the "war of the others" because of the role that Israel, Palestine, Jordan, and Syria played in shaping its contours, this war of wars "dissolved into a self-perpetuating miasma of violence that saw countrymen, formerly allied groups, and families turn on each other."[3] It is still more complicated. These "other" communities were themselves "born" from the same Motherland through the 1916 Sykes-Picot Agreement and the 1920 Balfour Declaration. The catastrophic result was a region that is colonially divided and ruled against itself. As such, the very postcolonial struggles to transcend these historical agreements lead to endemic destruction and subjugation. The civil war was, at many different levels, a family affair. Mouawad explained in a 2008 interview that the interpersonal silence about the war was so strong because "it was a shameful war, where fathers killed sons, where sons killed their brothers, where sons raped their mothers. . . ."[4] This is reproduced in the region's geopolitical drama. *Scorched* explores the cycles of war, loss, pain, trauma, rape, death, and torture.

Nawal's story also draws from particular events and individuals in Lebanon's official history. Nawal's character is modeled on the actions of Soha Bechara who, at twenty-one, attempted to assassinate Antoine Lahad, the patriarch of the South Lebanese Army, by infiltrating his inner family circle. She shot him two times but failed to kill him. In return Bechara served ten years in Khiam prison—notorious for its torture, rape, and brutality. Mouawad explained, "for ten years she heard the crying and pain of the tortured. To try not to become mad, she began to sing."[5] She was known as the *woman who sings*. Furthermore, Nawal's encounter with the burning bus in the play (earning it the title *Scorched*) stands in for the event that is said to have sparked the civil war on April 13, 1975, when Christian militants attacked a bus of Palestinian refugees (known as the Ain El Remmaneh massacre).[6] A cycle of violent reprisals would follow the incident sinking the country deeper and deeper into an all-consuming war.

Scorched uses this all-consuming, cannibalistic clash of different communities to speak, in excess of the historical incidents of war, to the very conditions of Middle-Eastern politics. Nawal's communities (self, family, religion, nation, and globe) constitute the existential cartography of Janine and Simon's journey and also serve as the explicit barriers for their peaceful coexistence. Not only were maternal connections lost but families were also turned inside-out. Conflict led to warring factions and militant heroics. It gave birth to alliances, nations, and countries. It also produced all-consuming anger, mistrust, and destruction. In this context, we understand that Janine and Simon remain at war with the world, their mother, and each other. Like this geopolitical womb, they came into the world red-faced, screaming about the injustice of their historical conditions.

The play resonates with the audience at the level of this assumed cartographic imaginary. Perhaps our warring pasts condemn our future generations, ideals, and lives to cycles of blood, hatred, conflict, and injustice? Accepting *Scorched* as a cartography of conflicting communities leads to particular kinds of modern liberal geopolitical lamentations: Why can't *we* just get along? Can't *we* just forget and start anew? How does one reconcile historical tensions between contemporary communities? How does one cultivate tolerance between differences? How does one establish peace between those who hate or want to kill each other? Whither redistribution? What of reparations? Why do they hate us? The tragedy of world politics is that these liberal lamentations reproduce the very problems that they desire to solve and create a feeling of hopelessness. They fuel the very catastrophic cycles of violence that they desire to solve because they do not adequately engage the very conditions of community that give rise to the problems in the first place. These liberal laments fail us. At best they lead to tragedy and provoke social, moral, and political reflexivity.

However, *Scorched* is also more than a tragedy. It does not only accept and reproduce the above cartographic imaginary in its finale in order to provoke reflexivity. Instead, *Scorched* shifts how to think about the drama of international conflict and security. Although it too posits that communities are condemned to reproduce the past in the present and the future, it is not their essential differences that condemn them. It does not accept the well-rehearsed geopolitical imaginary of clashing communities. Instead, communities are condemned because they are organized around immunity mechanisms that simultaneously enable and threaten their existence. Instead of perpetually ending with liberalism's warring communities, this chapter begins with warring immunities.

Immunity

Roberto Esposito argues that immunity and community are inseparable and endemic. This focus on inseparability is unusual because of the way in which community is typically assumed to be an organic and/or a given aspect of human life. For Esposito the problem resides in how those who are interested in politics think about community. He explains, "[P]olitical philosophy tends to think of community as a 'wider subjectivity'" (*CM*, 2). Community, it is assumed, belongs "to subjects that join them together . . . and qualifies them as belonging to the same totality" (*CM*, 2). Community, it is assumed, is produced when individuals join or merge to produce a greater totality. As such, when we think together we become Lebanese, Italian, or Canadian. Esposito rejects this organic widening from subject, self, family, pack, tribe, nation, and globe assumed in modern politics. He explains,

> Community cannot be thought of as a body, as a corporation in which individuals are founded in a larger individual. Neither is community to be interpreted as a mutual, intersubjective "recognition" in which individuals are reflected in each other so as to confirm their initial identity; as a collective bond that comes at a certain point to connect individuals that were before separate. (*CM*, 7)

Instead, Esposito takes an etymological approach to alternatively show that the root of community is *munus*: an obligation, debt, or gift that cannot be repaid. "*Communitas*," Esposito explains, "is the totality of persons united not by a 'property' but precisely by an obligation or a debt; not by an 'addition' but by a 'subtraction': by a lack, a limit that is configured as an onus" (*CM*, 6). Community is not the sum total of individuals; it is a debt.

Immunity means that one is not subject to the obligation or debt of community. Immunity derives from the absence of a *munus* or office (*CM*, 4). Without office, someone is not indebted or responsible and is therefore entitled to privileges that can interrupt the social fields in which they travel. In diplomatic immunity, for example, because the diplomat has no office in a particular country, they are not obliged or held responsible to the host country's rules. However, immunity is not the opposite of community. On the contrary, the logic of immunity is embedded in what it means to be a community. Community is the ephemeral result of the immunitary negotiation with this debt. The result is a perpetual drive for protection, survival, and/or control of the community.

As such, immunity is not the same as security and securing community. Whereas the logic of security promises the end to violence, danger, and risk through expulsion or exclusion, the logic of immunity does not and cannot offer such a promise. Instead of vanquishing the enemy, the immunitary logic brings the danger and risk into the existential character of the community. Esposito explains, "Evil must be thwarted but not by keeping it at a distance from one's borders; rather, it is included inside them. . . . The body defeats a poison not by expelling it outside the organism, but by making it somehow part of the body" (*IM*, 8). Inoculation does not happen after a community has come into existence and been exposed to a danger; community is itself the expression of having been immunized. In sum, "community is simply the interface of its own immune system, the margin without depth along which immunity folds self-reflexively back on itself" (*IM*, 51). Community is an absence manifest by a particular way of thinking about health, life, and vitality (i.e., immunity).

Esposito explains this dynamic of immunity-community in terms of an included exclusion and/or excluded inclusion. Each is a negation of a negation. The very thing that is said to be outside is central to maintaining the mechanism that keeps it "outside." To be vaccinated, for example, requires incorporating that from which immunity is sought. Your immunity does not depend on your character; it depends on what you have internalized. In Esposito's words, "the immunitary logic is based more on a non-negation, on the negation of a negation, than on an affirmation" (*IM*, 8). Instead of affirming the qualities present in an already existing community, "immunity constitutes or reconstitutes community precisely by negating it" (*IM*, 9). The result is that to be protected against violence, the community must wield violence. To guard against lack, the community must mobilize scarcity. To deal with risk, the community is endangered. To confront illness, we must become infected. As such, immunity "can prolong life, but only by continuously giving it a taste of death" (*IM*, 9).

An equilibrium kind of tolerance or affirmative biopolitics is generated between community and immunity in its positive form. Immunity to immunization develops. Immunity and community become indistinguishable. Esposito gives the example of motherhood to show that instead of the fetus being the product of foreign invasion, it is transformed into life. He explains that "what allows the child to be preserved by the mother is not their resemblance but rather their diversity transmitted hereditarily from the father" (*IM*, 170). Instead of rejecting that which is foreign, this tolerant immunity to immunization, transforms the foreign into something new. In other words, "the equilibrium of the immune system is not the result of defensive mobilization against something other than the self, but the joining line, or the point of convergence, between two diverse series" (*IM*, 174). In this way the mother nurtures the fetus in the womb through an immunitary mechanism. Esposito explains: "The mother is pitted against the child and the child against the mother, and yet what results from this conflict is the spark of life. Contrary to the metaphor of a fight to the death, kill or be killed, what takes place in the mother's womb is a fight 'to life'" (*IM*, 171). The child is not an extension or an advisory of the mother; the child is something new.

In its negative form, Esposito also warns that immunity can get out of hand and drive toward thanatopolitics. Instead of generating a kind of tolerance that assures more life, immunity can turn on community and drive toward self-annihilation. He explains that "[w]hat threatens to catastrophically break down is precisely the immune apparatus that turns the negative into a positive, since at this point it identifies with the negative precisely in order to absorb it dialectally" (*IM*, 108). Esposito highlights the thanatopolitical danger: "instead of adapting the protection to the actual level of risk, it tends to adapt the perception of risk to the growing need for protection—making protection itself one of the major risks" (*IM*, 16). In other words, it produces an overreaction to defend itself from its defences (like an anaphylactic reaction) and "immunity slips into the vortex of infinite duplication" (*IM*, 108–09). The very mechanisms that protect, in turn, become the enemy to be destroyed. This condition is biopolitical, not geopolitical. Life here is condemned to live in the midst of the immunity mechanisms that simultaneously enable and threaten its existence. Biopolitics proper takes life as its object and reduces life to its "state of absolute immediacy" (*IM*, 14). Organized around the permanence of death and sickness, life itself is pushed to the margins. Biopolitics forces life to live more by continually exposing it to death and sickness. "Life is sacrificed to the preservation of life" (*CM*, 14). If death and sickness become the permanent and exclusive horizon of life, we become the living dead or deadly alive. Threat, danger, terror, risk, etc. become the permanent norms. The result is that communities are always at war with themselves because of the immunitary mechanism that founded them and now sustains their existence. While they are intertwined, runaway immunity is the prevailing force. Catastrophe.

Esposito's wider project seeks to exceed this catastrophic tendency. He asks, "How are we to fight the immunization of life without making it do death's work? How are we to break down the wall of the individual (community) while at the same time saving the singular gift that the individual (community) carries?" (*CM*, 19). To this end, I turn to a detailed reading of Mouawad's *Scorched* in order to shift away from thinking in terms of geopolitical conflicts. I mobilize Esposito's proposition that we live in immunities and that those immunity logics exist in complex relationships with each other. On this re-reading the potential for lived transformation that Esposito's wider project gestures toward can be developed. This push also shows the limits of Esposito's project, which relies on the biological to heal instead of seeking lived political transformation.

Warring Immunities

Nawal's life is exemplary of the immunity paradox that Esposito identifies. The very mechanisms designed to protect become the dangers that threaten her existence. However, Nawal's life is not a single, uniform expression of the modern immunity paradox. Instead of being caught in international relations, we witness the assembled dramas of her warring immunities. When taken together, they exceed any one community. Warring immunities express different political registers and are organized around a plurality of different modalities and performances. It is this plurality, I argue, that allows for political transformation. Whereas Esposito explores the immunity paradoxes associated with fear, guilt, and hope in the philosophical projects of Hobbes, Rousseau, and Kant, I show that each immunity paradox is differently positioned with respect to time and that each temporal dimension is depicted in different character's performances. This temporal emphasis gives more room to seek political transformation than Esposito's canonical architecture. The three immunity paradoxes in *Scorched* are: (1) Living in the Past, (2) Living in the Present, and (3) Living in the Future.

Living in the Past is an immunitary logic. Nawal declares in her will and testament that, "childhood is a knife stuck in the throat. It can't easily be removed" (*SC*, 6). We create the past because it promises to safeguard a sense of community in the present. Living in the past is instrumental to the survival of the nation, clan, family, etc. A taste of these memories and gestures is given in order to protect and reproduce the present community. However, living in the past offers a constant diet of future conflicts, recalcitrant practices, and delusions of grandeur. Community is produced by responding to past debts via immunity. Having been warned about living in the past, history remains a national past-time.

When Nawal's mother prohibits her from keeping her baby (Nihad) because the father (Wahab) was a Muslim refugee from Palestine and not of their community, religion, or soil, we are assured that tradition protects the integrity of these genealogical lines. When Nawal resists the idea of abandoning Nihad to the orphanage, her mother, recalling a decision she had to make for herself in the past, gives Nawal an exceptional option:

> Then you will have to choose. Keep this child and this instant, this very instant, you will take off those clothes that don't belong to you and leave this house, leave your family, your village, your mountains, your sky and your stars, and leave me. . . . (*SC*, 19)

In the end, the logic of the past wins. With happiness and marriage prohibited, the baby is born out of wedlock, sent to an orphanage, and abandoned to history. Paradoxically, buried memories and community amnesia simultaneously hide and fuel intergenerational and interfamilial anger. That which protects community also endangers it. Foreshadowing impending civil war, Nawal's family connections are twisted, territories are betrayed, and established values become liabilities. Nawal gives up Nihad, leaving no historical trace except for *a little red clown nose* that she sneaks into the baby's bundle.

Later we encounter Nihad who, like his estranged mother, is caught in the immunity paradox: the past condemns at the same time as it defines him. In a "retelling of ancient times," Chamseddine—the leader of a warring faction explains how he found a young Nihad wandering the ruins of his community, looking for meaning in his life. The war gave him meaning: Nihad fought for the cause, for the people, and for the refugees. The audience witnesses Nihad's diabolical skill as a sniper. In war, even a ruthless killer can be championed as a great protectorate of the innocent. Nihad was no simple defender of the weak, however. Nihad was living in a melancholic slumber; he was obsessed with photographing the people he killed because, as images and memories, they were more alive to him when they were dead. Nihad became more diabolical—often singing while sniping. One day, he just upped and left the rebels. Nihad was obsessed with finding a past life that had abandoned him. Nihad's ruthless ability derived from having been sent to the orphanage. Nihad kills in the present in order to reclaim what he lost in his past. Killing gives him meaning. Sniping is his life form. This obsession gives his killing a ruthless quality, an artistic genius, and the clarity of a machine that Chamseddine used, paradoxically, to protect his people and homeland.

Yet, the immunitary danger of such protection exists in its reversibility. Chamseddine explains how Nihad was turned when he was captured by the invading army: "They didn't kill him. They kept him and trained him. They gave him work" (*SC*, 76). Living in the past defines what present can be lived. Nihad's quest was reversible but not changeable. He was given new work in Khiam prison—notorious for its torture, rape, and brutality. Nihad's sense of belonging was defined by past relationships of guilt, obligation, and debt living. To live in the past required a constant violence—a community vigilance—that kills in the name of the living. Like his grandmother and his mother, he searched for the past in order to secure the present.

Living in the Present is also connected to a parallel sense of immunitary obligation. Instead of guilt, the present's debt is reinforced by fear. Living in a state of fear is a normal state of affairs for those who champion the present. Fear becomes an obligation and a debt to the present. Esposito explains that, while the promise of self-preservation generates the kind of hope for peaceful coexistence—hope is only fear's cousin (*CM*, 22). Explaining the importance of Thomas Hobbes to the modern immunity paradigm, he states, "fear does not disappear, however. It is reduced but it does not recede. Fear is never forgotten . . . fear is part of us; it is we outside ourselves. It is the other from us that constitutes us as subjects infinitely divided from ourselves" (*CM*, 23). The dynamic unfolds as solutions: in order to escape an originary and indeterminate fear, we accept and institute a regularized amount of fear in a covenant thereby normalizing the experience as a present and stabile order (*CM*, 24–25). Fear is normalized through an ever-present vigilance called anger. Those who live for the moment only do so in the perpetual fear of losing it. If there is to be hope, then there must be fear. The immunization that Hobbes proposes removes these uncertain fears and institutes new vigilant relationships between the individual and the community (i.e., the State).

After having given up her son as her mother insisted, Nawal's grandmother (Nazira) warns Nawal to forgive and forget. Nawal is counselled to take up the enabling tools of modern, civil education and live *her* life in the *now*. Her grandmother says,

> for you, it's just beginning. . . . We . . . our family, the women in our family . . . are caught in the web of anger. We have been for ages: I was angry at my mother, and your mother is

> angry at me, just as you are angry at your mother. And your legacy to your daughter will be anger too. We have to break the thread. So learn to read, learn to write, learn to count, learn to speak. Learn. Then leave. (*SC*, 22)

Nawal does leave. She learns to read, write, count, speak, think, forgive, and forget. But the web of anger—the warring immunities—surpasses her family relations. The immunities exist in the broader sectarian violence that crisscrosses the region. Driven by her search for her son, Nihad, Nawal dives into the civil war in Lebanon. The character Sawda, Nawal's sidekick, manifests the fear and anger of war. In war nothing is past and nothing is future—everything is now. Before committing herself to the assassination of Chad, the ruthless Paramilitary Patriarch, Nawal explains her motivation to Sawda,

> NAWAL: I made a promise . . . I promised an old woman I would learn to read, write and to speak so I could escape poverty and hatred. And this promise is going to guide me. No matter what. Never hate anyone, never, ...
>
> SAWDA: So what can we do?
>
> . . .
>
> NAWAL: Listen. We are going to strike. But we are going to strike a single spot. Just one. And we're going to hurt. We won't touch a single man, woman or child, except for one man. Just one. We'll get him. Maybe we'll kill him, maybe we won't, that doesn't matter, but we'll get him. (*SC*, 54, 55)

Nawal's hope of settling the score through one single violent act is the present promise to erase a past, create a future, and put an end to the cycle of war. It is exceptional. It appears that in repayment of a debt an exceptional sacrifice and extraordinary action is required of Nawal in the present. She infiltrates the patriarch's lair; but she fails to kill him. Her action only creates new debts and obligations within her overlapping warning immunities. In describing the persistent climate of revenge, guilt, and fear in war, the doctor at the orphanage (whose quote also opens this paper) answers Sawda's question: Why did the refugees take the children?

> *Doctor*: Out of revenge. Two days ago, the militia hanged three refugees who strayed outside the camps. Why did the militia hang the three refugees? Because two refugees from the camp had raped and killed a girl from the village of Kfar Mamira. Why did they rape the girl? Because the militia had stoned a family of refugees. Why did the militia stone them? Because the refugees had burned down a house near the hill where the thyme grows. Why did they burn down the house? . . . There must be a reason, that's as far as my memory goes, I can't retrace it any further, but the story can go on forever, one thing leading to another, from anger to anger, from sadness to grief, from rape to murder, back to the beginning of time. (*SC*, 35)

While this appears to be a story about the past influencing the present (as in Living in the Past), it is instead an immunity reproduction of the present in terms of anger and fear over a perceived past.

This cycle of anger and fear reemerges in Nawal's son Simon to whom she also bequeathed a letter. Simon is defined by his anger and fear. Simon is a boxer and he only wants to focus on his immediate match (*SC*, 41). However, he is not a very good fighter. He blames his failure on his conditioning; however, his handsome coach (Ralph) explains it otherwise:

> RALPH: You're never going to qualify if this keeps up. Put on your gloves and I'll tell you what your problem is.
>
> . . .
>
> RALPH: You're not looking! You're blind! You don't see the footwork of the guy in front of you. You don't see his defence . . . That's what we call a peripheral vision problem. (*SC*, 12)

Simon is nothing but immediate reaction. He has no perspective, no foresight, no peripheral vision, and no art. When Janine phones to relay important information about Nawal having been jailed, tortured, and raped, Simon responds, "I'm not interested in knowing who she was! I'm not interested! I know who I am today, and that's enough for me. . . ." (*SC*, 59). He lives in the present; he reacts with anger and fear to everything that comes before him. He is blinded by his burning emotions. He cannot tolerate the past and yet he perpetually fails to secure his future.

Although the grandmother (Nazira) warns Nawal about the legacy of anger to be left to her daughter, it is Simon who embodies the blind rage about which she warns. His present anger is simultaneously a strategy for dealing with the hurt of the past and the anxiety about the future. It is an immunity strategy that condemns Simon, like Nawal, Nihad, Lebanon . . . to perpetually reenact the very conditions that require more and more anger and anxiety.

Living in the Future is an immunitary logic of hope. Preparation, planning, preemption, and resilience are all logics of a future that simultaneously promise safety for a community. Yet, living in the future also generates blind spots and misdirected efforts that can lead to danger and/or downfall. When a future is imagined and then acted upon as if it were a certainty, that future replaces the pluripotentiality of other unknown futures to come. A community secures a future but, unable to see the alternatives, becomes vulnerable. Esposito explores this problematic in the form of Law (*CM*, 65). Developed in relationship to Kant's categorical imperative, the law is the fullest promise of obedience to the ideal. He explains that the law is not the will birthed at the origin of the community; instead the "law establishes the community, which in turn constitutes those areas in which the law is applicable" (*CM*, 65). From the future, the law gives itself to itself. As such, "free will, or freedom (that which is perceived in living in the present) thus becomes the same principle by which and from which, simultaneously, freedom itself originates: the origin of its own origin, an origin that is (in)original, having originated from that which originates. . . ." (*CM*, 67). The idea of law, therefore, originates from an assumed future that enacts itself in the present on behalf of the future. Yet, it necessitates itself because it also makes the community vulnerable to unknown futures (and unintended consequences).

This kind of logical abstraction is seen in Janine's dilemma. Unlike Simon's emotional present, Janine lives in an abstract future. She is a PhD student in mathematics and studies graph theory and the problem of polygons. She explains to her students that the mathematics that she studies is "totally different (than common math) since we will be dealing with insoluble problems that always lead to other problems, every bit [as] insoluble" (*SC*, 13). The problems are unsolvable because they derive from ideal forms (polygons) yet are to be used to describe real, lived, present situations. Squaring the distance (in a graph) between the ideal and the real in any language is a task in which "you'll never succeed. All graph theory is essentially based in this problem which remains for the time being impossible to solve. And it's this impossibility that is beautiful. The mystery of the Polygon remains intact" (*SC*, 13).

As though a mathematical puzzle, Janine seeks hope in her mother's silence and her bequeathed quest to find her father. This is her chance to square the ideal of the family with her lived experience. She sets off to discover the truth—the law of family. However, as Janine's maths predict, her study only leads to new and unexpected problems. This squaring of the circle does not allow for unexpected and implausible futures. Janine discovers that after the attempted assassination of Chad, Nawal was taken to Kfar Rayat. There Nawal developed a futurist strategy of resilience whereby she sang to overcome/endure/negate/frame the rape, torture, and brutality present in the prison. She sang to drown out the sounds of torture, to keep her sanity, and to offer the other inmates hope (*SC*, 6). Her commitment to a future ideal self—the woman who will sing in the face of horror—made it impossible for her jailer to break her. Although he violated her body, he could not touch her spirit. Her strength resided in a future self that categorically transcended her present horrors.

True to the immunity paradox, this strength ironically also made her vulnerable. Her torturer, Abou Tarek, was a twisted and sadistic man. He saw his torture, rape, and killing as a form of art. As such, he was drawn to this strong woman who sang and, because he liked her voice and strength, he did not kill her. She saved a bit of herself so that, like a resilient seed, she would one day grow healthy again to see justice and that this day of justice would redeem a past tortured life lived in the future tense. She stood as a testament, a promise to the future beyond war. Yet, the strength of her future self who sings in the face of torture condemned her present self to endure Abou Tarek's immediate passions. She endured ten torturous years of bare existence in the promise of a future to come.

Nawal eventually gets to see the future promise of justice when indicting Abou Tarek at the war crimes tribunal.

> *Nawal*: The woman who sings. Now you remember. You know the truth of your anger towards me, when you hanged me by the feet . . . the shards under my fingernails . . . the gun loaded with blanks against my temple . . . the gunshots and death that are part of torture, and the urine on my body, yours, in my

> mouth, on my sex, and your sex in my sex, once, twice, three times, so often that time was shattered. My belly was growing big with you, your ghastly torture in my belly, and left alone, all alone, you insisted that [I] be left alone to give birth. Two Children. Twins. (*SC*, 63)

Abou Tarek's ultimate torture was to make the woman who sings give birth to the children of her enemy, the children of her torturer, the children of her rapist. Abou Tarek is Janine and Simon's mathematical father. Not just for the ten years in the prison, however, Nawal would endure his torture for life in and through her/his children. Her resilience gave birth to survival but not to a life she could love. She explained,

> You made it impossible for me to love my children. Because of you, I struggled to raise them in grief and in silence. How could I tell them about you, tell them about their father, tell them the truth which, in this case, was a green fruit that could never ripen? (*SC*, 63)

His future/her future? Nawal did not predict this future; Janine's quest for the truth could not imagine it possible. Biological birth could not resolve political tensions in a fight for life. Nawal believed that justice would be done if she survived another day. She believed in the natality of law's covenant. And here is the gamble of resistance defined as resilience and/or attrition: one hopes that the sacrifice made in daily doses—sometimes at a level that negates life itself—will one future day allow justice, law, and life to be victorious.

> I promise you that sooner or later they will come and stand before you, in your cell, and you will be alone with them, just as I was alone with them, and like me, you will lose all sense of being alive. A rock would feel more alive than you. (*SC*, 63)

And then, the full weight of tragedy and mathematical impossibility . . . After her searing indictment Abou Tarek pulls out *a little red clown nose* that his mother had snuck into his blanket when she abandoned him to the orphanage. Suddenly, Nawal's future does not appear, the past makes no sense and the present becomes sublime. These modes of apprehension fail to make sense of what was promised. The sovereign covenant meets the monster at the door. Simon discovers the catastrophe that is stuck in Nawal's throat.

> *Chamseddine to Simon (aka Sarwane)*: The earth stops turning. Sarwane, it stops turning. Yes, that's right, he tortured your mother, and your mother was tortured by her son and the son raped his mother. The son is the father of his brother and his sister. . . . Inside you, Sarwane, the silence of the stars. And your mother's silence. Inside you. (*SC*, 77)

Political Transformation

Esposito posits that motherhood, as affirmative biopolitics, can negate the slide toward biopolitics, the reduction of life to its measure and its repatriation in thanatopolitics. In motherhood there is the possibility of the immunization of immunity. Instead of rejecting that which is foreign, Esposito argues, this tolerant immunity to immunization transforms the foreign into something new. The mother nurtures the fetus in the womb through an immunitary mechanism that negates the spiral towards the biopolitical/thanatopolitical continuum of Kill or Be Killed, Kill to Live More, Live More to Kill, and Kill and be Killed.[7] Esposito argues that what takes place in the mother's womb is a "fight to life" (*IM*, 171). As such, the child is not an extension or an advisory of the mother; the child is new life.

So what happened with Nihad/Abou Tarek? What happened with Simon and Janine? What happened to Lebanon? Why could biological birth, the fight to life, not overcome the terror of Lebanon's civil war and Nawal's rape? What lead to the corruption of biological motherhood and affirmative biopolitics? How could biological motherhood drive toward death, annihilation, civil war, brutality, and silence?

Esposito posits, drawing from Freud, that the incest taboo is the first biopolitical law.[8] Nihad is born from geopolitical incest in the region and Simon and Janine are born from biopolitical incest between mother and son. The second biopolitical taboo is cannibalism—consumption of that which is like you. The civil war consumes brother, sister, cousins, mother, father, neighbor, and national alike. As Sawda, Nawal's sidekick, explains: "They say 'the war won't catch up with us' I answer, 'yes, it will. The earth is being devoured by a red wolf' " (*SC*, 29). The biological process cannot negate incest or cannibalism; incest and cannibalism invokes another immunitary mechanism that drives toward death. As such, biological time and biological birth will not heal all wounds. Geopolitical and biopolitical wounds will grow deeper with each successive generation. In a world turned inward, the thanatopolitical grows with each cycle of violence. The impossibly of life will grow darker each day. . . . In Lebanon, Syria, Palestine, Israel. . . .

With the biopolitical/thanatopolitical catastrophe revealed at the indictment, Nawal abandons the fight for biological survival and accepts her silence and death. She gives up on fighting for any past, present, or future life. She has exhausted all efforts. Exceeding the dark side of biopolitics cannot be left to biological time, to the unfolding of life itself. Indeed, and this is where we depart from Esposito, Mouawad's *Scorched* makes clear that transformation must be political—not biological.

In her exhaustion Nawal makes her final gesture. Instead of repaying the debt connected to her survival, she makes the relations that have defined her life into a new political problem to be solved; a problem that cannot be solved by past, present, or future immunity. Instead of reproducing catastrophe, she becomes political. Becoming political requires that she return to the problem of her life by reencountering the actual life that she has lived (via the letters) and opening it up to indeterminacy. To this end, she draws from the temporal pluralities inherent in her warring immunities. Instead of relying further on past, present, or future immunities, Nawal draws upon the indeterminate temporality of two political events: the past-future and the future-past. She seizes the past in order to counter-actualize the future and the future in order to dramatize the past.

Whereas catastrophe reproduces the very enabling conditions that created the problem in the first place, an event embodies general transformation. An event here is not a moment, a sudden action or a happening. An event is an intensification of relations that forges new connections, interactions, and crystallizations. Gilles Deleuze explains, "Events are like crystals, they become and grow only out of the edges, or on the edge."[9] As such, it is not helpful to talk about the past, present, or future of events. "Events are never causes of one another, but rather enter the relations of quasi-causality, an unreal and ghostly causality, endlessly reappearing in the two senses (of past and future)."[10] In other words, an event betrays the immunological and epidemiological sense of causal time that gives us past, present, and future. An event amplifies from past-futures and/or future-pasts.

On the one hand, the past-future is when a future is brought into existence in and through a promise. Deleuze gives the example of an actor. The actor's present is always simultaneously a future and a past. He explains that "the actor maintains himself (*sic*) in the instant in order to act out something perpetually anticipated and delayed, hoped for and recalled. . . . The actor thus actualizes the event . . . the actor delimits the original, disengages from its abstract line, and keeps from the event only its contours and its splendour, becoming thereby the actor of one's own events—a counter-actualization."[11] What Nawal creates is a quasi-cause—"a counter to the actual, but a virtual that is not chaotic."[12] She becomes worthy of the event in and through the past-future of the letters she bequeaths.

Through the letters, Nawal makes Simon and Janine political actors. Without knowing, the twins are made to renegotiate the conditions and relations of their and Nawal's actual existence. As they follow Nawal's script, these ghostly letters, caught between past and future, evoke and rework the *dramatization* of Nawal's life. Dramatization restages the relationships that are involved in having produced the actual. Gilles Deleuze explains that *dramatizations* "are dynamisms, dynamic spatial-temporal determinations, that are pre-qualitative and pre-extensive, taking 'place' in intensive systems where differences are distributed at different depths, whose 'patients' are larval subjects and whose 'function' is to actualize Ideas. . . ."[13] Ideas for Deleuze are problems that incite the philosophical drive to create concepts that are adequate to the event. Explaining this process, Iain Mackenzie and Robert Porter remark, "the dramatization of concepts, therefore, requires setting them (i.e., concepts) into relationships with each other in ways that express the intensity of the relationships they already express; it is a way of determining the Idea that the concept expresses."[14] "An idea" they explain, "is the real problem that makes us think conceptually."[15] As such, through the course of the play Simon and Janine become counter-actualizations of Nawal's life. They are not overwhelmed by the catastrophic revelations (in the way they unknowingly were before receiving the letters). Because of their dramatization they become different actors. They are no longer defined by the actual lives they had been condemned to live in the present and live in the future. They have become new people born from horrible circumstances.

On the other hand, it is not enough to be open (or opened) to change. In addition to the past-future that allows for counter-actualization, the future-past shapes the dramatization. A new debt, a new idea must be made and a new belief in the world must be found. To paraphrase Mackenzie and Porter, through dramatization a new concept is *brought to life*.[16] The future-past recovers a debt, promise, or idea in its indeterminacy. A future-past is a renegotiation between what is to come and what has been forgotten. The future is not given in the past and yet nothing has really disappeared.

Nawal explicitly recovers the red clown nose that, to use Benjamin's phrase, "flashes up at the instant it can be recognized."[17] After all the years Nawal and Nihad spent together, it is only in and through the clown nose that she recognizes her loved son in the lived life of Abou Tarek. It is the artifact from the past that she seizes to open toward a transformed future. Nihad's horror becomes the object of Nawal's love. Why?

Nawal returned and remade the problem of her life. She returned to the idea of her life that she posed as a problem when she was forced to abandon her baby and chose community. She had promised Nihad "No matter what happens, I will always love you." This is a promise that we are expected to make to our life, our world, and our family. But she is incapable of loving Nihad because of her past, present, or future ability to forgive and forget. She has broken her promise and this is the problem of her life. However, because she remembers her love for Wahab, Nihad's father, she can transform the future-past of Simon and Janine via the letters. She creates new life beyond catastrophe. She does not forgive; she dramatizes her counter-actualization. Wahab's phrase "whenever we are together, everything is better" becomes her new promise and debt—she accepts her silence and death.

> WAHAB: Listen to me, Nawal. I don't have much time. At dawn, they're taking me away, far from here and far from you. I've just come back from the rock where the white trees stand. I said goodbye to the scene of my childhood, and my childhood is full of you, Nawal. Tonight, childhood is a knife they've stuck in my throat. Now I'll always have the taste of your blood in my mouth. I wanted to tell you that. I wanted to tell you that tonight, my heart is full of love, it's going to explode. Everyone keeps telling me I love you too much. But I don't know what that means, to love too much, I don't know what it means to be far from you, what it means not to have you with me. I will have to learn to live without you. Now I understand what you were trying to say when you asked: "where will we be

> in fifty years?" I don't know. But wherever I am, you will be there. . . . No matter where I am, we will be together. There is nothing more beautiful than being together. (*SC*, 20)

Nawal ceases to fight for her life upon seeing the clown nose. She stops trying to resolve her warring immunities and instead she accepts a new drama that, for her, ends where it already began. The result is that Simon and Janine need not reproduce the warring immunities that defined her life. They can fight to live anew.

A Dramatization of Catastrophe

We confuse the nature of *Scorched* if we think that the play's catastrophe occurs at the moment that Nawal realizes that her son is her historical rapist and the biological father of her children. Nawal's life was not catastrophic because of a perpetual clash of overlapping communities. It is catastrophic because she could not renegotiate the plurality of debts imposed on her by competing immunities. She could not exceed the very logics that sought to protect her and solve the problem of her life. In her penultimate gesture, she remade a promise to the world; through drama she lives beyond her actual life. The catastrophe predates Nawal's life.

Nawal's life is a testament to the contemporary catastrophic condition. Although the audience experiences her catastrophe in the second to last scene, the play has provided a counter-narrative from the beginning. Nawal was always already dead; Greater Palestine is already in ruins. With Simon and Janine, we are renavigating and rewriting her life. She has already lived through her rape, war, loss, trauma, birth, etc. The retelling of her life prepares us to realize that catastrophe cannot be deferred. Although we make incredible efforts to fend off impending catastrophe, *Scorched* teaches that we are already in the midst of catastrophe: endemic violence, poverty, alienation, exclusion, exploitation, destruction, extinction, terror . . . The catastrophe is already upon us. The catastrophe is the disappearance of the past, present and future. As such, the problem is not fending off that (next attack, downturn, spill, disaster, accident, etc.) which is coming toward us from the past, present, or future; it is about transforming the political conditions of our lives so that it becomes possible, not simply to live or survive, but to be born again.

Contrary to the promise of Esposito's immunity/community logic, affirmative biopolitics is insufficient to deal with the pain, trauma, anger, and hatred seen in *Scorched*. Relying on biology and its cadre of forces (emergence, evolution, and resilience) is insufficient. Instead, it is the political transformation of survival into something else that must be highlighted. Like Nawal, who offers Simon and Janine the letters, Mouawad, as a playwright offers us the problem. He makes theatre a problem; not for management, reflection, or survival, but for living. He re-dramatizes life beyond survival. Through the amplification of contingencies (living in the past, present, future, future-past, and past-future) we can make something new out of our catastrophe.

It is the reading of Nawal's will and testament, a transformative and political promise, which casts Simon and Janine as actors in their own lives. It is because of the letters and the will's instructions that Simon and Janine can dramatize and counter-actualize the catastrophe of Nawal's geopolitical life and renegotiate the actuality of their own existence. In discovering Nawal's new problem and promise, they can transform their own lives. This is political, not biological, rebirth. The political is not opposed to the catastrophic. The political is the counter-actualization and dramatization that allows us to live in and through catastrophe.

Notes

1. Wajdi Mouawad, *Scorched*, trans. Linda Gaboriau (Toronto: Playwrights Canada Press, 2005). Hereafter parenthetically cited as (*SC*).

2. This paper examines the core mechanisms of the catastrophe and as such does not protect secrets of the play.

3. Emily Hoffman, "A Brief History of the Lebanese Civil War," in American Conservatory Theater, *Words On Plays: Insight Into The Play, The Playwright, and The Production* Scorched (American Conservatory Theater, 2012), 22. Available at https://www.act-sf.org/content/dam/act/education_department/words_on_plays/Scorched%20Words%20on%20Plays%20(2012).pdf (accessed September 30, 2014).

4. Quoted in Kate Goldstein, "A Life of Resistance: A Brief Biography of Soha Bechara," in American Conservatory Theater, *Words On Plays: Insight Into The Play, The Playwright, and The Production* Scorched (American Conservatory Theater, 2012), 31. Available at https://www.act-sf.org/content/dam/act/education_department/words_on_plays/Scorched%20Words%20on%20Plays%20(2012).pdf (accessed September 30, 2014).

5. Quoted in Dan Rubin, "Wajdi Mouawad: At Home with Words," in American Conservatory Theater, *Words On Plays: Insight Into The Play, The Playwright, and The Production* Scorched (American Conservatory Theater, 2012), 6. Available at https://www.act-sf.org/content/dam/act/education_department/words_on_plays/Scorched%20Words%20on%20Plays%20(2012).pdf (accessed September 30, 2014).

6. Hoffman, "A Brief History of the Lebanese Civil War."

7. Geoffrey Whitehall, "The Biopolitical Aesthetic: Toward a Post-Biopolitical Subject," *Critical Studies on Security* 1, no. 2 (2013): 189–203.

8. Kiarina Kordela, "Biopolitics: From Supplement to Immanence: In Dialogue with Roberto Espositio's Trilogy: *Communitas, Immunitas, Bios*," *Cultural Critique* 85 (Fall 2013): 8.

9. Gilles Deleuze, *Logic of Sense* (New York: Columbia University Press, 1990), 9.

10. Ibid., 33.

11. Ibid., 151.

12. Gilles Deleuze and Felix Guattari, *What is Philosophy* (New York: Columbia University Press, 1996), 156.

13. Gilles Deleuze, *Dessert Islands* (New York: Semiotext(e), 2004), 108.

14. Iain Mackenzie and Robert Porter, "Dramatization as Method in Political Theory," *Contemporary Political Theory* 10, no. 4 (2011): 490.

15. Ibid., 490.

16. Ibid., 483.

17. Walter Benjamin, "Theses on the Philosophy of History," in *Illuminations*, ed. Hannah Arendt, trans. Harry Zohn (New York: Schocken Books, 1968), 255.

12

Becoming Normative

Law, Life, and the Possibility of an Affirmative Biopolitics

Patrick Hanafin

> la norma giuridica non si regola sulla persona, bensi si genera nella vita in comune.[1]

Introduction

In April 2014, the Italian Constitutional Court ruled that the prohibition on the use of donor eggs and sperm in IVF treatment contained in the 2004 Assisted Reproduction Act was unconstitutional.[2] This was the 29th challenge to the Act's constitutionality in its ten years in existence, and the 20th successful one. This case completes the judicial unwriting of the legislation achieved by constant legal challenges by groups and individuals affected by the Act's prohibitions. The law on assisted reproduction is a clear example of a normativization of life which ordered citizens ability to make choices in relation to assisted reproduction and gave symbolic legal recognition to the embryo. Most of the Act's provisions have now been declared unconstitutional as well as being contrary to the European Convention of Human Rights and Fundamental Freedoms in a Grand Chamber decision of 2013.

The 2004 Italian Assisted Reproduction Act, Legge 19 febbraio 2004, n.40, "*Norme in materia di procreazione medicalmente assistita*" displays an auto-immunitarian logic in which law is used to protect a particular national narrative based on Roman Catholic heteropatriarchal family values. The Act set out to create a model which protected Life in the abstract and restricted the ability of living citizens to make free decisions. It prohibited the testing of embryos for research purposes, embryo freezing, pre-implantation genetic diagnosis for the detection of genetically transmitted diseases, donor insemination, denied access to assisted reproduction services to single women, and provided that no more than three ova be fertilized *in vitro*, and that these be transferred to the womb simultaneously. Once couples agreed on the treatment they would not be allowed to withdraw their consent. Any medical professional attempting to carry out procedures prohibited by the legislation would face prison terms or fines, as well as suspension from the medical register. The law directly contradicted the provision in Article 31 (2) of the Constitution of the Italian Republic, which states that no protection independent of the mother shall be accorded to the unborn. Indeed, the Constitutional Court has held that the welfare of the embryo or fetus does not override a woman's right to health. The Act limited access to in-vitro fertilization to those categorized as infertile or sterile couples. Couples who were not so defined but who were carriers of a hereditary genetic condition would not have access to assisted reproductive services. The Act limited access to assisted reproductive services to adult heterosexual couples who are either married or in a stable relationship, are of a potentially fertile age, and are both living. As Ingrid Meltzer has observed, the law led to the construction of the embryo as "a new citizen subject."[3]

The Act epitomizes a negative biopolitics which orders the lives of citizens and makes them the objects of the norm, i.e., normativizes life. The Act instantiates what Roberto Esposito would call a "politics over life." Citizen contestation of the Act, on the other hand, has led to a judicial unwriting of its prohibitions. Here we have a parallel creation of law which is also a hollowing out of the law from within, a case of vitalizing an abstract norm. Citizens have inhabited the law, have taken up residence there, changing it from within. Citizen resistance to the Act demonstrates that the biopolitical imperative to control lives is not a one-way street and can be resisted by the micropolitical acts of citizens. It is in Esposito's terms an instantiation of a "politics of life" which points to the possibility of an "affirmative biopolitics" (*II*, 185). In other words, it is a politics which does not valorize an abstract ideologically rigid notion of Life that restricts individual lives, but performs a politics of life which is driven by actions of individual living beings acting in relation with one another. As

Esposito observes: "essa passa per la disattivazione dei dispositive autoimmunitari e per l'allargamento dello spazio del commune" (*II*, 185). In such an affirmative biopolitics we see the move from what Adalgiso Amendola has called "the absolute normativation of life . . . [to] . . . the vitalization of the norm."[4] What can be seen in such moments of contestation is the emergence of what Amendola terms: "a law that isn't imposed on subjects or that creates them artificially . . . but that might be produced together with the same processes of subjectivation through which subjectivities as such are formed."[5] For Amendola: "This affirmative biojuridical moment . . . can open to a reading of the same juridical normativizaton as a process that isn't superimposed abstractly over conflicts. . . . Therefore, a path can be opened, a bumpy one admittedly, though not impassable, from a *law of the immune* to *a law of the common*."[6]

Law's Politics Over Life

The Italian law on assisted reproduction reveals a bio-theopolitics in which life becomes the ground of a conflict between competing models of community, one immunitarian, and based on rigid delineations of a national essence, and the other plural, singular, and never completed or formed. In the Italian case the Act is the manifestation of a theo-political discourse which valorizes "Life" as an abstract value. In this politics what is sought is the emplacement of a particular religious traditionalist thinking at the heart of secular political life. The rhetoric of embryo politics endows the embryo with the characteristics of a fully-fledged citizen. In such a politics the embryo represents the promise of regeneration and perpetual life. The rhetorical linking between "embryo" and "Life" becomes part of the fantasy narrative that the pro-life movement fashions. As Condit has noted in relation to the pro-life politics of anti-abortion campaigners, but which can be applied *mutatis mutandis* to the politics of embryo protection:

> [T]he major rhetorical effort of the pro-life movement was . . . expended in constructing . . . the verbal linkages between the terms *fetus* and *Life*. The concrete term *fetus* and the abstract value of Life were woven together primarily through a frequent recitation of the claim that the authority of "science" had discovered that the fetus was a human being from the time of conception.[7]

In pro-life rhetoric doctors who conduct pregnancy terminations are constructed as having transgressed the "natural" order of things and are somehow tainted or associated with death and murder. As Merton has put it:

> [T]he movement accuses anyone who condones legal abortion of, at the very least, standing by and doing nothing while millions of innocent human beings are slaughtered. The logic goes this way: a) zygotes/embryos/fetuses are human beings in the fullest sense of the term, and therefore deserving of protection; b) abortion kills zygotes/embryos/fetuses; therefore c) abortion is murder, and d) anyone who condones abortion condones murder.[8]

Such a politics over life constructs women who exercise freedom of reproductive choice as enemies of the fetus. This discourse homologizes the fetus and the nation into a figure faced with death from a threatening force, i.e., self-determining women. As Lauren Berlant has noted in relation to the pro-life narration of national identity:

> [T]he normativity of pro-life society dictates that once pregnant the woman loses her feminine gender, becoming primarily a mother . . . In protecting the fetus from the woman they divide into a nongenital "female" part—the maternal womb, which really belongs to the fetus—and a potentially malevolent section, composed of a sexual body (un)governed by a woman's pseudosovereign consciousness.[9]

This immunitarian model of the maternal-fetal relation makes women invisible, creating a model of community in which the mother is seen as somehow a threat to the fetus. This model fails to see the contingent and relational element of the maternal-fetal dyad and exemplifies in Derrida's words: "a state power where sovereignty is itself essentially phantasmatico-theological and, like all sovereignty, is marked by the right of life and death over the citizen, by the power of deciding, laying down the law."[10]

Such an immunitarian logic sets up a conflictual relationship between the embryo seen as worthy of legal protection and the mother seen as threatening to the nation. The 2004 Assisted Reproduction Act was driven by the need to reclaim a conservative notion of the nation based on patriarchal family values. This is a politics which lessens the freedom of living citizens in the interest of an abstract notion of the sanctity of life. As Esposito has put it: "il concetto . . . di 'sacerta' della vita e spesso usato come un dispositivo di esclusione o di soppressione di altre vite, giudicate non altretanto rilevanti" (*II*, 185). Precisely in this sense we see the use of the term sanctity of life and right to life as non-negotiable and as exclusionary *dispositifs*. The lives and beings of individuals denied access to reproductive choice and justice are disregarded along with the constitutional rights to health and self-determination in the name of an abstract signifier of Life as normativizing and subject forming. The model incorporated in the Act sees the creation of a notion of community as immunity against intruders and ultimately against death, a social compact built on the desire to survive. As Krause and De Zordo have noted in this regard:

> The rigid politics of life operating in Italy supported by the Catholic Church and sympathetic politicians defends the 'life' and the rights of the embryo and the ideal Catholic family at all costs. As a result, women who do not have children or who postpone motherhood are stigmatized, as are infertile women and couples who confront a restrictive law on medically assisted technologies, which excluded single women and same-sex couples.[11]

By valorizing life as true abstract life, woman is relegated to the status of mere life, an intermediary figure used as a means of reproducing life.

Vitalizing the Norm: Toward a Politics of Life

In this regard what the Italian case allows us to see is the operation of a biopolitics which both governs and excludes. This exclusionary consequence of biopolitics has been well defined by Didier Fassin as "about inequalities in life which we could call bio-inequalities."[12] Such a notion of bio-inequality includes a "withholding [of] recognition from the other."[13] It is precisely this withholding of recognition from individuals denied reproductive choice that has led to a counter-politics of resistance against the legislation. This resistant biopolitics of living citizens calls for a continuous struggle to maintain and win rights. It allows us to move from "a rigid politics of life" to a "power of life as such."[14] It demonstrates the power of individuals acting in concert to contest draconian state action and allows us to see in Fassin's terms that "another politics of life is possible."[15] Ingrid Meltzer has described such individuals as biological citizens: "Speaking in the name of their physical vulnerability and mobilizing their damaged bodies, they acted as 'biological citizens.'"[16] Such biological citizens use their bodies as a strategic means of achieving reproductive freedom and choice. The notion of the biological citizen is an interesting one in that it brings together both the reality of contemporary political regimes in which we are all the subjects of governance, with the coexisting ability to resist such governance in the mode of an affirmative biopolitics. It creates a space of resistance in which citizens take on an active role in contesting the manner in which their citizenship is constructed. Here what we witness is a vitalizing of the norm as a form of counter-conduct. What we have witnessed in the last ten years of contestation of the 2004 Act is the slow and painful process of the becoming sovereign of what the law has considered to be mere life.

In order to contest their construction by the law as citizens without reproductive rights, patients' rights groups affected by the legislation engage in, in Stephen Collier and Andrew Lakoff's terms, a "counter-politics of sheer life," which they define as "a claim to state resources that is articulated by individuals and collectivities in terms of their needs as living beings."[17] This praxis of active citizen resistance to claim new rights or to reclaim rights taken from them is an instance of an affirmative biopolitics which opens up the field for political resistance by those categorized as bare life or lives excluded from human rights protection. In Joao Biehl's terms, we can see such patient action as enacting a form of "biocommunity."[18] This community is made up of: "a . . . group of . . . patients [which] fights the denial of rights and carves out the means to access them empirically."[19] Such a biocommunity proposes an alternative model of community, not one based on auto-immunity, but based on what Esposito would call a model of "common immunity" (*IM*, 165–77). Unlike the model of auto-immunity, the idea of "common immunity" provides a mode of imagining and experiencing political community as open and contestatory. This model of common immunity is based on an understanding of community not as bounded and based on defending the nation from enemies, but on an inclusion of one within the other, on community as difference (*IM*, 165–77). This undoes the symbolic conservative notion of the self-sufficient nation under attack from others seen as enemies.

Esposito thinks political community in terms of difference and not in terms of a hegemonic identity. Esposito is concerned with the deactivation of *dispositifs* such as that of the juridical person in favor of the expansion of the space of the common. For him, the notion of the person inherited from Roman law constitutes a mode of disembodying the individual, of devaluing the material in favor of the abstract. This valorization of an abstract rational subject over the mere life of actually existing individuals leads to the denigration of the flesh in favor of abstract reason. Individual lives are devalued in favor of a politics which rules over these lives. As Esposito observes: "The person doesn't coincide with the body in which it inheres, just as the mask is never one with the actor's face."[20] As such it can be seen as a *dispositif* which allows the material lives of individuals to be subjected to power, to be constructed as objects of power. For Esposito, the idea of material singularities coming together to act in common provides an example of the power of a collective being in common of material lives. In effect it provides a means of countering the *dispositif* of the person and the biopolitical governance of lives through law. It allows us to see how an impersonal force may dissolve the enforced distinction between *bios* and *zoé*, between the homogeneous nondifferentiated subject of power and the flesh of individual lives, and exposes singularity in difference. This overcoming of the separation between bare life and Life is for Esposito the task of an affirmative biopolitics, which is a continuous task, a work in progress, a becoming, and a continual beginning.[21] For him: "An affirmative biopolitics always involves decisions about life, its meaning, its different demands, its preservation, and its expansion."[22] Esposito leaves as an open question how we think a relationship between humanity and rights which is: "freed from the subjective slant of the legal person and brought back to the singular, impersonal being of community" (*TP*, 140). He notes, however, that such a relationship is "only conceivable starting from life" (*TP*, 140). As such this relation of humanity to rights can only be based on the actions of living beings. In what follows, I attempt to outline how such a relationship of the human to rights might manifest itself by analyzing the contestation of the 2004 Assisted Reproduction Act by the "biological citizens" affected by its prohibitions.

Life, Humanity, Rights

One of the most important in the long series of legal challenges by individuals affected by the 2004 Assisted Reproduction Act was the decision of the Regional Administrative Tribunal of the region of Lazio of January 21, 2008. This case was initiated by the World Association for Reproductive

Medicine (WARM), a not-for-profit organization which represents the interests of professionals working in the area of medically assisted reproduction. The action challenged, *inter alia*, the legitimacy of the Code of Practice introduced by Ministerial Decree in 2004, as being *ultra vires* the powers of the Minister of Health, as well as the constitutionality of Article 13 (which prohibited embryo experimentation), and Article 14 (which provided for the transfer of no more than three embryos to the womb simultaneously) of the 2004 Act. WARM also contested the conflation of the terms sterility and infertility in the Act and the legal status accorded to the embryo in the Act. This challenge, which also had the support of a number of other reproductive rights organizations, was opposed by the Italian government together with a number of conservative civil society organizations, such as the Movement for Life, who intervened *ad opponendum*. The Court in its decision overruled parts of the Code of Practice introduced pursuant to the 2004 Act in July 2004. The impugned provisions related to Article 13.5 of the Act which prohibits experimentation on human embryos. The decision also raised doubts over the constitutionality of Article 14.2 of the Act. In effect, what the decision did was to overrule the limitation on pre-implantation genetic diagnosis of embryos for observational purposes only, on the basis that such a provision could not be enacted by delegated legislation. The Minister of Health had therefore exceeded his powers in introducing this measure by ministerial regulations. As a result of this decision, the guidelines on assisted reproduction were revised on April 11, 2008, to remove the *ultra vires* limitation on pre-implantation genetic diagnosis for observational purposes only.

In its decision, the Lazio court also referred the question of the constitutionality of Article 14 of the Act to the Constitutional Court. In its decision of April 1, 2009 the Constitutional Court reversed the prohibition contained in Article 14 of the 2004 Act on the transfer in any one cycle of a maximum of no more than three embryos. In addition to the referral from Lazio, the Court also received two referrals from the Tribunale Ordinario of Florence from its decisions of July 12 and August 12, 2008. In both of these decisions the Florence court questioned the constitutionality of Article 14 of the Act insofar as it prohibited the freezing of spare embryos, the imposition of a maximum limit of three embryos which could be created in any IVF treatment cycle, and the need for their simultaneous transfer to the patient's womb. In addition the court questioned the constitutionality of Article 6 (3) of the Act which decreed that once a woman had consented to the simultaneous transfer of these three embryos she could not withdraw that consent. The Constitutional Court in its decision held that Article 14.2 of the Act was unconstitutional and in particular breached Article 3 of the Constitution in relation to equality and Article 32 of the Constitution which upholds the right to health. As a result of this decision, Article 14.2 of the 2004 Act was no longer to be interpreted as placing a limit on the number of embryos to be transferred. The Court held that the number of embryos transferred in any treatment cycle should be based on individual medical opinion based on the facts of each patient's case. The decision also overruled the ban in Article 14 (1) on the freezing of embryos. As a result of the decision, embryos which might not be used in a treatment cycle may now be frozen. The Court, in referring to Article 1 of the Act, noted that the interests of all parties (not just the embryo) should be considered citing the Constitutional Court's previous jurisprudence on abortion in which the rights of the woman to self-determination and health should be given priority.

Following the Constitutional Court ruling there have been a number of subsequent successful challenges to the Act in the lower courts. The Tribunale Civile of Florence in its decision of October 6, 2010, overturned the ban on IVF with donor eggs or donor sperm in Article 4 of the Act and referred this aspect of the Act to the Constitutional Court for review. On October 21, 2010, the Tribunale Civile of Catania made a similar ruling, questioning the constitutionality of the ban on IVF using donor gametes. In the decision of the Tribunale Civile of Salerno of October 13, 2010, the limitation in Article 1 of the 2004 Act on access to in-vitro fertilization to those categorized as infertile or sterile was successfully challenged. The Court ruled in favor of access to pre-implantation genetic diagnosis in the case of a couple who were neither sterile nor infertile. The couple suffered from amyotrophy, which causes the progressive wasting of muscle tissues.

The 2004 Act was the object of a further important Constitutional Court decision in May 2012. This case concerned the question of the prohibition of IVF using donor gametes under Article 4 of the Act. The decision however turned out to be more of a non-decision in that it held that the cases should be referred back to the regional courts from which they issued for rehearing. The case involved references from three lower courts, in Florence, Catania, and Milan in relation to Article 4, paragraph 3 of the Act (which bans IVF using donor gametes), on the grounds of potential constitutional incompatibility. The Florence case involved a couple, SB and FB. The male partner was infertile and the couple required access to donor sperm. The clinic which they attended could not carry this out as the Act prevented it from doing so. The court was of the opinion that the impugned section of the Act was, *prima facie*, unconstitutional but noted that it needed to refer the matter to the Constitutional Court as lower court judges do not have the power to declare a part or whole of a statute unconstitutional.

The reference from the court in Catania concerned a couple, PC and GR. PC suffered from premature menopause and attended a clinic in order to request an egg donation. However she was prevented from doing this by the prohibition contained in Article 4, paragraph 3 of the 2004 Act. The Court noted a *prima facie* breach of the Constitution and noted that such a procedure was medically necessary. Again, due to the inability of lower court judges to declare statutes unconstitutional, the case was referred to the Constitutional Court. In the Milan case a couple, EP and MM, required sperm donation as the male partner suffered from azoospermia. In this case the prohibition contained in Article 4, paragraph 3 of the 2004 Act prevented the couple from gaining access to such a procedure. All three courts noted that there was a *prima facie* constitutional violation. The justification given by lawyers on behalf of the Government in the argument before the Constitutional Court for such a prohibition was the right of the child to know the biological identity of his or her parents. This justification had more to do with a conservative mentality in relation to family relations rather than any rights of the child involved.

On hearing the references before it, the Constitutional Court decided not to decide and instead referred the issue back to the lower courts. The Constitutional Court used as a justification for this the then-recent decision of the Grand Chamber of the European Court of Human Rights in the case of *S.H. and Others v Austria* (Application No. 57813/00, Grand Chamber decision November 3, 2011). In that case the Grand Chamber held that there was no violation of Article 8 of the European Convention of Human Rights and Fundamental Freedoms in a case involving a challenge to the provision of the Austrian Assisted Procreation Act which prohibits the use of sperm from a donor for IVF and ova donation in general. The Austrian Assisted Procreation Act only allows IVF with gametes from the couples involved. Even though the Grand Chamber noted that there was a clear trend across Europe in favor of allowing gamete donation for IVF, it added that an emerging consensus was still under development and so was not as yet based on settled legal principles. The Court held by a majority of thirteen votes to four that there had been no violation of the Convention. The Grand Chamber further noted that the Austrian legislation was not disproportionate as it had not banned individuals from going overseas for infertility treatment unavailable in Austria. This assumes, without thinking, that couples are in a position to engage in such reproductive tourism.

The decision of the Grand Chamber was entirely at odds with the First Instance ruling in the same case on April 1, 2010, *S.H. and Others v Austria* (chamber judgment, April 1, 2010), which held that the impugned section of the Austrian legislation breached Article 8 of the European Convention of Human Rights and Fundamental Freedoms as this prohibition interfered with the couple's right to access treatment which would allow them to found a family. The lower courts had noted, based on the First Instance decision of April 1, 2010, in *S.H. and Others v Austria*, that the prohibition in the 2004 Act of IVF using donor gametes constituted a breach of Articles 8 and 14 of the European Convention of Human Rights and Fundamental Freedoms. The Constitutional Court observed that as the Grand Chamber had overruled this decision the referring courts should rehear these cases based on this new development.

Since the Constitutional Court judgment in May 2012, the Court of First Instance of the European Court of Human Rights handed down a decision against Italy in relation to the prohibition of pre-implantation genetic diagnosis where a couple are carriers of a genetically inherited condition. In the case of *Costa and Pavan v Italy* (Application No. 54270/10), a couple, Mr. Pavan and Ms. Costa, both carriers of a hereditary illness, cystic fibrosis, wished to prevent this condition being inherited by any second or subsequent child they might have together. In September 2006 they gave birth to a child with cystic fibrosis, only then becoming aware that they were both carriers of the disease. The couple have a one in four chance of having a child born with the condition and a one in two chance that any future child of theirs will be a carrier of the condition. They want to ensure that any further child they have will neither have nor be a carrier of the condition. The 2004 Act prevents access to pre-implantation genetic diagnosis to couples suffering inherited genetic conditions. It only allows access to screening for infertile couples or where the male partner has a viral disease which can be transmitted through sexual intercourse, such as HIV, or Hepatitis B or C. Since these exceptions did not apply to this couple, the only option open to them as the law stood was to have an abortion on discovery via fetal testing that the future child was either a sufferer or carrier of the condition. In fact, Ms. Costa conceived a child with cystic fibrosis and decided to undergo an abortion in February 2010.

In their application to the European Court of Human Rights in Strasbourg, the couple relied on Article 8 in conjunction with Article 14 of the *European Convention on Human Rights and Fundamental Freedoms*. Their complaint was that their right to privacy and family life had been infringed in that they were not allowed access to pre-implantation genetic diagnosis to allow them to prevent the birth of a child with cystic fibrosis. They also claimed that they suffered discrimination compared to infertile couples or those couples in which the male partner has a sexually transmitted disease. In its decision of August 28, 2012, the Court of First Instance of the European Court of Human Rights held unanimously that the ban on access to pre-implantation genetic diagnosis for couples with genetically inherited diseases infringed Article 8 of the Convention. The Court found that there was no breach of Article 14. The Court held that the desire of the couple to have a child who was not affected by a genetically inherited disease of which they were healthy carriers and to undergo pre-implantation genetic diagnosis and IVF in order to do so was protected by Article 8 as it formed part of their right to private and family life (*Costa and Pavan v Italy* [Application No. 54270/10]. Court of First Instance decision 28 August 2012, para. 57). The Court unanimously declared that the 2004 legislation was incoherent in that on the one hand it prohibited the transfer of only embryos which were not affected by cystic fibrosis and on the other hand it allowed the couple to abort a fetus affected by this condition. There was a clear impact on the couple's Article 8 rights in this case as a result.

The Court distinguished the decision of the Grand Chamber in *S.H and Others v Austria* in which the Court allowed a wide margin of appreciation to Austria in legislating in this area. The Court concluded that the interference with the applicants' right to privacy constituted by the ban in the 2004 Act of pre-implantation genetic diagnosis to such couples was not proportional. The Italian Government entered an appeal against this decision. However, the Grand Chamber of the European Court of Human Rights rejected this appeal in February 2013, noting that the Italian Law on Assisted Reproduction was clearly incoherent and in breach of Article 8 of the *European Convention on Human Rights and Fundamental Freedoms*. The decision of the Court of First Instance of August 2012 is now the final word on the matter as far as the compatibility of the 2004 Act with the *European Convention on Human Rights and Fundamental Freedoms* is concerned. This decision strengthens the hand of those groups in Italy campaigning for the legislation to be reviewed.

The decision requires the Italian Government to revise the 2004 Act to make it compatible with the *European Convention on Human Rights and Fundamental Freedoms*. However, given the lack of willingness of successive Italian governments to move in this direction, it is unlikely that such a review process will begin immediately. What will continue to happen will be individual court challenges to the Act which will gradually have the cumulative effect of nullifying the Act's prohibitions. It will then be imperative even for unwilling politicians to act to introduce a law which is both coherent and compatible with the European Convention of Human Rights and Fundamental Freedoms.

This series of challenges is an example of the vitalization of the norm brought about by the dogged persistence of biological citizens. This active citizen politics allows us to see how the abstract control over Life exercised by the State in the name of religious ideology can be contested successfully. Such a mode of political intervention allows us to imagine a "politics of life" in the sense outlined by Esposito. Such a model stresses the need for continuous political engagement to make real the merely declaratory nature of rights. It is an active engagement with the promise contained in constitutional bills of rights to enable citizens to access rights in reality. As such, this recent episode in Italian political life has universal resonance in that it demonstrates clearly the need on the part of citizens to resist in contemporary regimes of biopower when their material lives are devalued and their full citizenship is threatened in the name of a totalizing narrative of Life. As Krause and De Zordo have put it: "the struggles around reproductive policies are articulated in juridical terms . . . and produce rights-bearing citizens pitted against each other . . . These new moral regimes generate social and political spaces for ongoing negotiation."[23]

Becoming Normative?

This example provokes us to think how the normativization of lives can be contested by the vital contestation of individual citizens. In this ongoing challenge to their normative ordering citizens engage in an active mode of using rights discourse in a subversive manner to undo accepted models of subjectivity, community, identity, law, and politics. Such a micropolitics of rights opens up the field for political resistance by those categorized as excluded from full citizenship. It is in Rosi Braidotti's terms a "move towards 'Life' as a non-essentialist brand of contemporary vitalism and as a complex system."[24] As such we come to see the emergence of a creative jurisprudence in the Deleuzean sense, a disarming of the normative ordering of individual lives. Such legal and political challenges are initiated by assemblages of individuals acting in concert to use rights discourse in a manner which would empower them. Such a creative bottom-up employment of rights as political weapons allows us to glimpse what Gilles Deleuze called the creative and collective praxis of jurisprudence. For Deleuze it is jurisprudence "that truly creates law/right."[25] Jurisprudence for Deleuze is not the abstract conceptulizaion of law or legal theory but rather an active mode of resisting established legal concepts and changing and troubling the law through the collective action of singularities. For Deleuze, it is not "established and codified rights that count, but everything that currently creates problems for the law and that threatens to call what is established into question."[26] In reflecting on Deleuze's thinking on and about law and human rights, Paul Patton observes that for Deleuze:

> jurisprudence was always a matter of politics, in the broad sense in which he understood the term . . . Jurisprudence involves the creation of new laws but also the creation of the rights that are expressed in these laws . . . in so far as jurisprudence is also a matter of politics, it involves the processes through which new ways of acting or being acted towards become established (or old ways disestablished).[27]

Esposito declares an affinity with Deleuze in his thinking which opens up a thinking of the impersonal and the anonymous, of a life rather than Life as abstract ordering. This thinking allows us to reframe the tired language of institutionalized and abstract subject-centered human rights discourse and instead forces us to think a materially embodied concept of rights as the collective action of singularities, post-human rights if you will. Such a praxis of rights is similar to what Paul Patton terms a "non-transcendent, immanent conception of rights."[28] These rights embody the claims of transversal assemblages of individuals who do not see a binary cut between thought and action, life and death, environment and humanity, or animality and humanity. Esposito's reliance on Deleuze is evident throughout his work and indeed we can see in his latest book, *Due: La macchina della teologia politica e il posto del pensiero* (2013), a suggestive opening to Deleuze's thought on jurisprudence in an affirmative biopolitical key when he observes in relation to Deleuze's notion of jurisprudence:

> For [Deleuze] . . . if the subject is no longer what s/he was because subjectivity is not measurable in temporal terms given that its past is always overtaken and reconfigured by the present, then it is no longer entirely subsumable by the order of law. That which its becoming something other escapes is precisely that normative regime which makes of the law not only a dispositive of guilt and punishment within a theo-political paradigm. This doesn't mean that Deleuze is placed outside or against the sphere of law. What he performs is a linguistic movement which moves the norm from the vertical relation of sovereign command to that of the horizontal plane of the form of life. The passage from one to the other, from the regime of imposition to that of metamorphosis is linked to the creative capacity of jurisprudence. If, in the *a priori* logic of law, facts like persons are subjected to an order which judges single cases on the basis of pre-existing principles, jurisprudence in the Deleuzian sense, conceived of in its originary creative power, derives principles from single cases.[29]

This overturning of our understanding of the established ordering of the norm allows the singular case to drive the norm, an instantiation of the vitalization of the norm, a vital resistance to the norm. Thus, what is at stake here is an understanding of jurisprudence as the possibility of making the norm something other than what it was, through the vital intervention of biological citizens.

We can see in the critical responsiveness to the 2004 Act the possibility of the creative transformation of this imposition of ordering. It provides us with a glimpse of a creative jurisprudence in which rights are not achieved by a top down "molarpolitics of public officials"[30] but come instead from the mobilization of self-styling selves, "the molecular movements of micropolitics."[31] In this example we witness the play between the *micropolitics* of movements of individuals who are attempting to self-style their reproductive choices, and the *molarpolitics* of politicians, who attempt to prevent the creation of this right. This *molarpolitics* is based on rigid moral beliefs and refuses to recognize contrary views. It blocks the dialogic political process and creates stasis. William Connolly has termed this behavior on the part of citizens an *ethos of engagement* with existing moral and social givens that may bring about unexpected consequences or transformations in the societal default thinking on bioethical issues. Similar to Esposito's notion of affirmative biopolitics, this active resistance on behalf of citizens affected by prohibitive legislation on bioethical issues leads to a reactivation of rights protection for those deprived of such protection. This process Connolly terms "an ambiguous *politics of becoming* by which a new entity is propelled into being out of injury, energy and difference."[32] Micropolitical movements such as the patients' rights groups in Italy who continue to call for a more and more liberal governance model for the assisted reproductive technology sector, or those individuals who bring legal challenges to the existing law, provoke us to rethink existing modes of addressing bioethical problems. Individuals take responsibility for themselves and work on the political and legal terrains to bring about real change. The intervention of individuals in the political scene via the creative use of jurisprudence allows precisely what cannot be brought about by appeals to politicians to change legislation, the move from, in Esposito's terms, the regime of imposition to that of metamorphosis.

Notes

1. Dario Gentili, *Italian theory: dall'operaismo alla biopolítica* (Bologna: Il Mulino, 2012), 217.

2. Andrea Gagliardi and Federico Mereta, "Fecondazione assistita, la Corte costituzionale boccia il divieto dell'eterologa in Italia," *Il Sole 24 Ore* (9 Aprile 2014), www.ilsole24ore.com.

3. Ingrid Meltzer, "Between Church and State: Stem Cells, Embryos, and Citizens in Italian Politics," in Sheila Jasanoff, ed, *Reframing Rights: Bioconstitutionalism in the Genetic Age* (Cambridge, MA: MIT Press, 2011), 105–24, 118.

4. Adalgiso Amendola, "The Law of the Living: Material for Hypothesizing the Biojuridical," *Law, Culture and the Humanities* 8 (2012): 102–18, 115.

5. Ibid., 115.

6. Ibid., 116.

7. Celeste Condit, *Decoding Abortion Rhetoric: Communicating Social Change* (Urbana: University of Illinois Press, 1990), 61.

8. Andrew H. Merton, *Enemies of Choice: The Right to Life Movement and Its Threat to Abortion* (Boston: Beacon Press, 1981), 7.

9. Lauren Berlant, *The Queen of America Goes to Washington City: Essays on Sex and Citizenship* (Durham: Duke University Press, 1997), 99.

10. Jacques Derrida, *The Death Penalty: Volume 1* (Chicago: University of Chicago Press, 2014), 5.

11. Elizabeth L. Krause and Silvia De Zordo, "Introduction: Ethnography and Biopolitics: Tracing 'Rationalities' of Reproduction across the North-South Divide," *Anthropology and Medicine* 19 (2012): 137–51, 143.

12. Didier Fassin, "Another Politics of Life is Possible," *Theory, Culture & Society* 26 (2009): 44–60, 49.

13. Ibid., 57.

14. Ibid., 49.

15. Ibid., 44.

16. Meltzer, "Between Church and State," 105–24, 111.

17. Stephen J. Collier and Andrew Lakoff, "On Regimes of Living," in Aihwa Ong and Stephen J. Collier, eds, *Global Assemblages: Technology, Politics, and Ethics as Anthropological Problems* (Oxford: Blackwell, 2005), 22–39, 29.

18. João Biehl, *Will To Live: AIDS Therapies and the Politics of Survival* (Princeton, NJ: Princeton University Press, 2007), 324.

19. Ibid., 324.

20. Roberto Esposito, "The *Dispositif* of the Person," *Law, Culture and the Humanities* 8 (2012): 17–30, 25.

21. See further, Miguel Vatter, "Biopolitics: From Surplus Value to Surplus Life," *Theory & Event* (2009): 12.

22. Roberto Esposito interviewed by Timothy Campbell and Federico Luisetti, "On Contemporary French and Italian Philosophy: An Interview with Roberto Esposito," *The Minnesota Review* 2010, no. 75 (2010): 109–18, 111.

23. Krause and De Zordo, "Introduction," 137–51, 148.

24. Rossi Braidotti, *The Posthuman* (Cambridge, UK; Malden, MA: Polity Press, 2013), 158.

25. Gilles Deleuze, *Negotiations: 1972–1990* (New York: Columbia University Press, 1995), 169.

26. Ibid., 153.

27. Paul Patton, "Immanence, Transcendence, and the Creation of Rights," in Laurent de Sutter and Kyle McGee, eds, *Deleuze and Law* (Edinburgh: Edinburgh University Press, 2012), 15–31, 20–21.

28. Ibid., 15.

29. Roberto Esposito, *Due: La macchina della teologia politica e il posto del pensiero* (Torino: Einaudi, 2013), 217.

30. William Connolly, *Why I Am Not A Secularist* (Minneapolis: University of Minnesota Press, 1999), 147.

31. Ibid., 149.

32. Ibid., 160.

Contributors

Alexander U. Bertland is an Associate Professor of Philosophy at Niagara University. His research has focused primarily on the philosophy of myth in Giambattista Vico. He is currently exploring how Vico's political philosophy and his account of the mythic origin of political institutions may inform contemporary analyses of community. This work is informed heavily by the work of contemporary philosophers Jean-Luc Nancy and Remo Bodei. He has also published in the area of business ethics.

Greg Bird is an Assistant Professor in the Department of Sociology, Wilfrid Laurier University, Canada. He is the coordinator of Techne: Wilfrid Laurier University Biopolitical Research Group and an international advisory board member for Workiteph: Network on Italian Thought and European Philosophies. His publications on contemporary Italian thought, philosophies of community, and biopolitics have appeared in English and Italian. He is the author of *Containing Community: From Political Economy to Ontology in Agamben, Esposito, and Nancy* (SUNY Press, 2016) and co-editor of *Community, Immunity and the Proper: Roberto Esposito* (Routledge, 2015).

Antonio Calcagno is Professor of Philosophy at King's University College, London, Canada. He is the author of *Giordano Bruno and the Logic of Coincidence* (1998), *Badiou and Derrida: Politics, Events and Their Time* (2007), *The Philosophy of Edith Stein* (2007), *Lived Experience from the Inside Out: Social and Political Philosophy in Edith Stein* (2014). He is also the editor of *Contemporary Italian Political Philosophy* (with SUNY Press).

Diane Enns is Professor of Philosophy at Ryerson University, Toronto, Canada. She is the author of three books: *Speaking of Freedom: Philosophy, Politics and the Struggle for Liberation* (2007), *The Violence of Victimhood*

(2012), *Love in the Dark: Philosophy by Another Name* (2016), and co-editor of *Thinking About Love: Essays in Contemporary Continental Philosophy* (2015). She is currently writing on community and loneliness.

Mark F. N. Franke is Associate Professor and Director of the Centre for Global Studies at Huron University College and is a core graduate faculty member in The Centre for the Study of Theory and Criticism at Western University, in London, Ontario, Canada. He is engaged primarily with research and writing on the politics of rights, movement and political ontology, and aesthetics of space, time, law, and geography. He returns, perpetually, to confrontations with problems in modern political theory and the ways in which they bring themselves to crisis, looking for moments of opportunity and optimism.

Federico Fridman is a Visiting Assistant Professor of Latin American Literature and Culture at Bucknell University. He received his PhD in Romance Studies from Cornell University in 2014, and holds a BA in Political Science with a concentration in political theory from the University of Buenos Aires. He has taught at the University of California–Riverside, and at the University of Buenos Aires's Law School and Political Science Department. He has published several articles and book chapters on Latin American literature, literary theory, political theory, critical theory, European continental philosophy, and transatlantic studies.

Patrick Hanafin is Professor of Law at Birkbeck Law School, University of London, where he also directs the Law School's Centre for Law and the Humanities. His research engages with questions of law and the biopolitical, law and literature, human rights and citizenship, and the construction of community and identity. He has held research fellowships at the European University Institute in Florence, at the Human Rights Program at Harvard Law School, and at the University of Cape Town. His books include: *After Cosmopolitanism* (with Rosi Braidotti and Bolette Blaagaard) (2013); *Deleuze and Law: Forensic Futures* (with Rosi Braidiotti and Claire Colebrook) (2009); and *Conceiving Life: Reproductive Politics and the Law in Contemporary Italy* (2007).

Federico Luisetti is an Italian philosopher and Professor of Italian Studies, Comparative Literature, and Communication at the University of North Carolina at Chapel Hill. He is the author of books and essays on critical theory, literary and visual studies, and political thought. He has edited, with John Pickles and Wilson Kaiser, the volume *The Anomie of the Earth: Philosophy, Politics, and Autonomy in Europe and the Americas* (Duke University Press, 2015). He is currently writing a monograph on the states of nature of the Anthropocene.

Alberto Moreiras is a Professor of Hispanic Studies at Texas A&M University. He is the author of *Interpretación y diferencia* (Madrid: Visor, 1991), *Tercer espacio: literatura y duelo en América Latina* (Santiago: ARCIS/Lom, 1999), *The Exhaustion of Difference: The Politics of Latin American Sultural Studies* (Durham, NC: Duke UP, 2002), *Línea de sombra: El no sujeto de lo político* (Santiago: Palinodia, 2008), and *Marranismo e inscripción, o el abandono de la conciencia desdichada* (Madrid: Escolar y Mayo, 2016). He is coeditor of *Journal of Spanish Cultural Studies, Res Publica: Revista de pensamiento político,* and *Política común: A Journal of Thought,* and of the University of Texas Press Book Series "Border Hispanisms."

Jonathan Short's work investigates problems in contemporary political theory, focusing on biopolitics, sovereignty and the nation. His published essays have examined the thought of Foucault, Agamben, and Esposito, as well as the political thought of the Frankfurt School. He is also co-editor of *Community, Immunity, and the Proper: Roberto Esposito* (Routledge 2015). Jonathan teaches in the Departments of Communication Studies and Social Science at York University in Toronto.

Inna Viriasova is a lecturer in the Department of Politics at Acadia University, Nova Scotia, Canada, where she teaches political theory. Her current research is in the areas of modern and contemporary political theory, critical refugee studies, non-Western political thought, and posthumanist security studies. She is the author of *At the Limits of the Political: Affect, Life, Things* (Rowman & Littlefield International, 2018).

Geoffrey Whitehall is a Professor in the Department of Politics and the Coordinator for the Social and Political Thought graduate program at Acadia University, Nova Scotia, Canada. His current research focuses on: Global Biopolitics; Sovereignty and Subjectivity; and Aesthetics and the Political Imaginary. For recent publications please see http://polisci.acadiau.ca/Dr._Geoffrey_Whitehall.html.

Index

affirmative biopolitics: Assisted Reproduction Act and, 242–247, 254–256; *Scorched* (Mouawad) and, 234–239

affirmative biopolitics—Esposito on: concept of, xiii–xiv, 165; deconstruction and, 88–96; the living person and, 102, 110–112; *munus* and, 57; resistance and, 71; the third person and, 81–82, 151, 194

Agamben, Giorgio: on archaeology of actuality, 162; on beatitude, 66–68, 82–83, 84, 98n28; on biopolitics, 48–49, 86–87; on community, xii, 35, 49, 51–55; on debt, 51–55, 58; Esposito and, 48–49, 55–56, 58, 61; Esposito on, 90–93; on ethics, 52, 54–55; on Heidegger, 83–88; on *homo sacer*, 175n30; Italian Thought and, 48, 88; on the person, 66–68, 70–71, 81, 82–88; on rights, 128; on virtualities, 166

Ain El Remmaneh massacre, 221

Alexander VI, Pope, 15

Amendola, Adalgiso, 243

animal belonging, 82

animism, 102, 112–121, 165, 171–173

animistic personhood, 113, 114–121

Antelo, Raúl, 195n5

anthropology, 73

the anti-political, 144, 160n30, 180, 181

Arendt, Hannah, xii, 74, 128–129, 134–136, 185, 196–197n18

Aristotle, 40, 129, 164–165

askesis, 136

Assisted Procreation Act (Austria), 250–251

Assisted Reproduction Act (Italy): affirmative biopolitics and, 242–247, 254–256; biopolitics and, 241–242; legal challenges to, 241, 247–253; normativization of lives and, 253–256

"El atroz redentor Lazarus Morell" ("The Dread Redeemer Lazarus Morell") (Borges), 187–188

Autobiography (Vico), 11–12, 22n9

Baer, Karl von, 84

Balfour Declaration (1920), 221

Balibar, Étienne, 39

bare life, 52, 55, 66, 83

Bataille, Georges, xii, 49, 92

beatitude: Agamben on, 66–68, 82–83, 84, 98n28; Deleuze on, 82, 84; Esposito on, 71, 82, 87

Bechara, Soha, 221

becoming-animal, xiv, 71, 102, 110, 111–112

Being and Time (Heidegger), 84

Being Singular Plural (Nancy), 49

Benjamin, Walter, 66, 98n28, 165, 166, 237
Bennett, Jane, 123n24, 171
Benveniste, Émile, xiv, 70, 78, 101–102
Bergson, Henri, 156, 164–165
Berlant, Lauren, 244
Betrachtungen eines Unpolitischen (Mann), 196n10
Bichat, Xavier, 71–73, 103–104, 106, 107
bioethics, 70, 75
bio-inequalities, 245–246
biopolitics: Agamben on, 48–49, 86–87; Assisted Reproduction Act and, 241–242; Deleuze on, 149–150; Esposito on, 48–49, 194, 225; Foucault on, xiii, 143, 146–150, 198–199n48; limits of, 79–80; origins of, 72–73; the person and, 104–107. *See also* affirmative biopolitics
bios, 101, 115, 118, 129, 247
Bios (Esposito), xii, 60, 155, 165, 194
Bioy Casáres, Adolfo, 191
Bird-David, Nurit, 115–116, 123n22
The Birth of Biopolitics (Foucault), 148–150
Blanchot, Maurice, xii, xiv, 81, 101–102
Bloch, Ernst, 166
the body, xiv–xv, 68–70, 162–165
Borges, Jorge Luis, 180–181, 187–194
Borgia, Cesare, 15
Bosteels, Bruno, 182
Braidotti, Rosi, 253–254
Bush, George W., 27
Bustos Domecq, H., 191
Byrd, Jodi A., 175n28

Cacciari, Massimo, xi, 88, 91–92, 179, 196n10
Campbell, Timothy, 51–52, 110, 143, 147–148, 157–158, 193
Canada, 127–128, 129
Canadian Spy Agency, 131
cannibalism, 234–235
Carillo, Gennaro, 3–4
catastrophe: Esposito on, 220, 225–226; *Scorched* (Mouawad) and, 219–220, 235–239
Categorie dell'impolitico (Esposito), xi–xii, 48–49, 90, 179–186, 193–194
Catholicism, 182–184, 241–242
Christian theology, 103–104, 107, 145–146, 162–164, 167
Cicero, 127
citizenship, 73, 139n3, 140n5
clinamen, 40
Collier, Stephen, 246
The Coming Community (Agamben), 49, 51–52, 55, 66
Communitas (Esposito): on biopolitics, 194; on community as impossible, 27; context of, 49; on etymological meaning of community, 32–35; on friendship, 165; on *munus*, xii, 151, 156; on nihilism, 43–44; on the proper, 59–60
community: Agamben on, xii, 35, 49, 51–55; Balibar on, 39; debate on, 49–51; desire for unity and, 28–29, 40–43; Gunsteren on, 39; Hobbes on, 204–206; Kant on, 206–209; Lingis on, 33; meaning of community life and, 28–29, 38–44; Nancy on, xii, 49, 93; Rancière on, 157–158; Stein on, 28–29, 41–42
community—Esposito on: Agamben and, 49; debt and, 56–59; definition and etymology of, xii, 28, 31–35, 39; dialectic of alienation and appropriation and, 49; Hobbes and, 204–206; as impossible, 27,

28; Kant and, 206–209; nihilism and, 28, 33–35, 43–44; sharing and, 50; war on terror and, 27, 30–31. *See also* immunity; *munus* (gift, obligation)
"Community, Immunity, Biopolitics" (Esposito), 57
Comte, Auguste, 72–73
The Concept of the Political (Schmitt), 182–183, 196n10
Condit, Celeste, 243–244
Confessions (Rousseau), 11–12
Connolly, William, 256
consensus, 157–158
Constitution of the Italian Republic, 242
Costa and Pavan v Italy, 251–252
Croce, Benedetto, 23n14, 24n20

Da fuori. Una filosofia per l'Europa (*From Outside: A Philosophy for Europe*) (Esposito), xv, 76, 88–94
Dardot, Pierre, 149
Darwin, Charles, 105
Dasein, 84–85, 156
death, 72, 104
death penalty, 29
debt, 51–60
De Constantia (Vico), 19–20
Deleuze, Gilles: Agamben and, 84; on animal belonging, 82; on beatitude, 84; on becoming-animal, xiv, 71, 102, 110, 111–112; on biopolitics, 149–150; on creative jurisprudence, 254; on deconstruction, 88; Esposito and, 70–71, 254–255; on *homo tantum*, 67–68; on the impersonal, 70–71, 81, 101–102, 109–110, 111–112, 134, 156; on life, 102, 110, 111–112; *Scorched* (Mouawad) and, 235–236
De l'évasion (Levinas), 81
De l'existence a l'existant (Levinas), 81
Derrida, Jacques: Agamben and, 84; on deconstruction, 89–90; Deleuze and, 67; Esposito and, 195n3; on "mere life," 96n1; on politics of separation, 95–96; on sovereignty, 244
Descartes, René, 6
Descola, Philippe, 105–106, 114, 123n22, 171
destituent power, 55, 61
De Zordo, Silvia, 245, 253
"Dialogue on the Philosophy to Come" (Esposito), 60
Dickens, Charles, 67
la differenza italiana, 47–48
Digest, 127
Dingpolitik, 170
Discourse on the Origin and Basis of Inequality Among Men (Rousseau), 6–7
Discourses (Machiavelli), 15
dissensus, 144, 153–155
divine providence, 8
dramatizations, 236–238
Driesch, Hans, 84
Due: La macchina della teologia politica e il posto del pensiero (Esposito), 5, 48–49, 254–256

economy of compensation (*economia del risarcimento*), 52
Empire (Hardt and Negri), 60
Engelhardt, Hugo, 75
Essay Concerning Human Understanding (Locke), 129–130
ethics: Agamben on, 52, 54–55; Esposito on, 58–59
ethos of engagement, 256
etymology: Agamben and, 53–54; Esposito and, 6, 28, 31–35, 39; Vico and, 6

Europe, xv
European Convention of Human Rights and Fundamental Freedoms, 250–253
European Court of Human Rights (ECHR), 250–253

factum, 6
fascism, 32–33, 41
Fassin, Didier, 245–246
Fernández, Macedonio, 187, 198n40
"La fiesta del monstruo" ("The Monster's Feast") (Bustos Domecq), 191
First Nations, 128
form-of-life, 55
fortune (*fortuna*), 16–17
Foucault, Michel: Agamben and, 84; on biopolitics, xiii, 143, 146–150, 198–199n48; on contemporary societies, 163; on deconstruction, 88; Esposito and, 70–71; on governmentality, 128, 130, 137; Hadot and, 136; on the impersonal, 70–71, 81, 101–102, 110–111; on life, 102, 110–111; on neoliberalism, 143, 146–150, 157; on outside, xiv; on the person, 128–129, 136; on resistance, 71, 110; on thanatopolitics, 210; translations by Esposito, xi; on the unthought, 110–111
French Revolution, 73, 74
French Theory, 48, 88, 89–90
Freud, Sigmund, 75, 234–235
friendship, 165
friendships, 40
The Fundamental Concepts of Metaphysics (Heidegger), 83–88

Gadamer, Hans-Georg, 7
Gaia theories, 171–172
Gaio, 162–163
Gentile, Dario, 47
German Philosophy, 48, 88, 89–90
Giarrizzo, Giuseppe, 24n20
Gilbert, Margaret, 139n3
Gnosticism, 98n28
Gouges, Olympe de, 154–155
governmentality, 104, 128, 130, 137
grand politics, 88–89
Grossman, David, 45n3
Guardini, Romano, 183
Guattari, Félix, 171
Gunsteren, Herman van, 39

Hadot, Pierre, 136
Hallowell, Irving, 123n22, 123n32
Hardt, Michael, 60
Harvey, Graham, 115, 123n22
Hegel, Georg Wilhelm Friedrich, 169
Heidegger, Martin: Agamben and, 83–88; Deleuze and, 67; on ex-sistence, 51; on the impersonal, 134; on "mere life," 96n1; Nancy and, 156; naturalism and, 169–170; on ontic guilt, 52–53
Heraclitus, 42
History of Sexuality (Foucault), 136
Hobbes, Thomas: immunity and, 226, 228; International Relations and, 202, 203–206; naturalism and, 169; on the person, 74, 163; on political categories, 194; Schmitt and, 183; Serres and, 172; on sovereignty, 107; Vico and, 3
Hölderin, Friedrich, 54–55
Homer, 9
Homo Sacer (Agamben), 48–49
homo tantum, 67–69, 82–83
honesty, 13–14
"Human Personality" (Weil), 198n43
human rights: Deleuze on, 254; Esposito on, 143, 144, 212–214;

Rancière on, 153–155; UDHR and, 73–74, 106, 128, 213
Husserl, Edmund, 67, 84, 139n3, 164–165

ideal eternal history, 7–10
ideology, 12–14, 15
"L'illusion comique" (Borges), 189–190
"Immanence: A Life . . ." (Deleuze), 67–68
immortality, 29
Immunitas (Esposito): on biopolitics, 194; on Carillo, 3–4; on "immunitary" paradigm, xii–xiv, 27, 150; on life and politics, 140n7; on private law, 57; on Vico, 21; on Weil, 58
immunity: Assisted Reproduction Act and, 241–242; neoliberalism and, 157–158; Tauber on, 35–36; as "tolerant immunity," 40–43
immunity—Esposito on: concept of, xii–xiv, 27–31, 159n23, 222–225; debt and, 56–59; effects of, 30–31; International Relations and, 209–212; motherhood and, 28, 35–38, 224–225, 234; the person and, 150; *Scorched* (Mouawad) and, 226–234; security and risk and, 131; as "tolerant immunity," 28–29, 35–38. *See also munus* (gift, obligation)
the impersonal: Deleuze on, 70–71, 81, 101–102, 109–110, 111–112, 134, 156; Foucault on, 70–71, 81, 101–102, 110–111; Heidegger on, 134; thanatopolitics and, 108; Weil on, xiv, 58, 76–77, 101–102, 193–194. *See also* the third person
the impolitical: Borges and, 180–181, 187–194; Bosteels on, 182; Esposito on, xi–xii, 59, 179–180, 181–186, 191–194; Lacoue-Labarthe on, 182; Nancy and, 182, 195n3; Schmitt on, 182–184, 185
"L'impolitico nietzscheano" (Cacciari), 179, 196n10
Improper Life (Campbell), 51–52
In Catastrophic Times (Stengers), 171
Incendies (film), 219
incest, 234–235
the infrapolitical (*lo infrapolítico*), 195n5
Ingold, Tim, 123n22
The Inoperative Community (Nancy), 49
inoperosità, 49
Institutiones (Gaio), 162–163
International Relations (IR)—Esposito on: Hobbes and, 202, 203–206; immunity and, 209–212; the impersonal and, 212–215; Kant and, 202, 203, 206–209, 215
Italian Thought, 47–48, 61, 88–96
iustitia aequatrix, 14
iustitia rectrix, 14

James, William, 172
Jankélévitch, Vladimir, xiv, 101–102
Jansenism, 11
Jhering, Rudolf von, 21
justice, 14
Justo, Agustín P., 197n26

Kafka, Franz, 187
Kant, Immanuel: Agamben on, 83–84
Kant, Immanuel—Esposito on: immunity and, 226; International Relations and, 202, 203, 206–209, 215; law and, 231
Karner, Dietrich, 122n19
Kierkegaard, Søren, 134
The Kingdom and the Glory (Agamben), 48–49

Kohn, Eduardo, 171
Kojève, Alexandre, xiv, 49, 82, 101–102
Krause, Elizabeth L., 245, 253

Lacoue-Labarthe, Philipe, 182, 195n3
Lahad, Antoine, 221
Lakoff, Andrew, 246
land rights, 128
Latour, Bruno, 114, 123n22, 162, 164–168, 170, 171–172
Laval, Christian, 149
Lebanese Civil War (1975–1990), 221
legal personality, 213–214
Leibniz, Gottfried Wilhelm von, 172
Leviathan (Hobbes), 163, 169, 203–206
Levinas, Emmanuel, xiv, 67, 78–81, 84, 101–102
liberal biopolitics, 106–107
liberal democracy, 73
"lifepower" (*Lebenskraft*), 41–42
Línea de sombra (Moreiras), 195n5
Lingis, Alphonso, 33
linguistics, 73
the living person (*la persona vivente*), xiv, 101–102, 109–113, 114–116, 118–121
Living Thought (Esposito), 4, 10, 47, 48, 56, 165
Locke, John, 3, 51, 129–130, 169
Lovelock, James, 171
Lyotard, Jean-Francois, 88

Machiavelli, Niccolò, 4–7, 10, 15–20, 43, 89
machinic animism, 171
Mackenzie, Iain, 236–237
Mann, Thomas, 196n10
The Man versus the State (Spencer), 198n40
Maori, 165
Margulis, Lynn, 171
Maritain, Jacques, 139n3
Marriot, McKim, 124n39
Marx, Karl, 29
Marxism, 10
Mauss, Marcel, 162, 165
McNay, Louis, 149
McPhail, Agnes, 128
melancholy, 75
Meltzer, Ingrid, 246
Merleau-Ponty, Maurice, 67, 164–165
Merton, Andrew H., 244
micropolitics, 255–256
molarpolitics, 255–256
monarchy, 20
La moneda de oro (The Golden Coin) (Borges), 192
mono-naturalism, 170
Moreiras, Alberto, 119, 195n5
Mosko, Mark, 117
motherhood, 28, 35–38, 224–225, 234, 244
Mouawad, Wajdi: . *See Scorched* (Mouawad)
Mounier, Emmanuel, 139n3
Mueller, Johannes, 84
munus (gift, obligation): concept of, xii–xiii, 6, 28, 32–35, 44, 223; debt and, 52, 56–59; deconstruction and, 93; the impersonal and, 143–144, 150–157, 158; Rancière's dissensus and, 153–155

Nagel, Thomas, 172
Nancy, Jean-Luc: on clinamen, 40; on community, xii, 49, 93; on debt, 53; Esposito and, 186; Esposito on, 95; Heidegger and, 156; on "withdrawal of the political," 182, 195n3
naturalism, 104–106, 112–113, 115–116, 161–162, 168–171
Nayaka, 115–116

Nazi ideology, 73–74, 79, 104, 106, 117, 135
Negri, Antonio, 47, 60, 93
neoliberalism: biopolitics and, 165; the economic and the political in, 50–51; immunity and, 157; the impolitical and, 179; the person and, 143, 147–150, 157–158
new animism, 102, 114–121
Ng, Hiu Lui, 97n21
Nietzsche, Friedrich, 84, 97n15, 134, 164–165
Nietzsche and Philosophy (Deleuze), 97n15
nihilism, 28, 33–35, 43–44

Ojibwa, 115
The Open (Agamben), 70, 82–88
operaismo, 49
The Order of Things (Foucault), 110–111
The Origins of Responsibility (Raffoul), 53
The Origins of Totalitarianism (Arendt), 134–136
Our Mutual Friend (Dickens), 67

panpsychism, 172
partage, 156
Patton, Paul, 254
Peirce, Charles Sanders, 171
Perón, Evita, 190–191
Perón, Juan Domingo, 189–191
the person: Agamben on, 66–68, 70–71, 81, 82–88; citizenship and, 139n3, 140n5; concept of, 129–130; debate on, 65–66; Deleuze on, 67–68; Foucault on, 128–129, 136; Hobbes on, 74, 163; Levinas on, 78–81; moral dimension of, 128–129, 135–136, 137–138; Taylor on, 136–137; Weil on, 76–77, 145, 151, 167; women and, 77–78. *See also* the impersonal
the person—Esposito on: Blanchot and, 81; concept of, xiv–xv, 129–131; critique of, 101–107, 134; genealogy of, 68–76, 127–128, 131, 143, 144–147, 162–165; governmentality and, 137; human rights and, 143, 144; immunity and, 150; the living person and, xiv, 101–102, 109–113, 114–116, 118–121; moral personhood and, 137–138; neoliberalism and, 143, 147–150, 157; Weil and, 76–77, 167. *See also* the third person
Le persone e le cose (*Persons and Things*) (Esposito): on affirmative biopolitics, 95–96; on the body, xiv–xv; on the impersonal, 76; naturalism and, 161–162, 168–171; on the person, 132–133, 162–165; on things, 165–172
Pettit, Philip, 139n3
philosophy of crisis, 15–18
Plato, 164–165
poetic wisdom, 8–9
police, 29–30
La politica e la storia (Esposito), 4, 7
political ontology, xi–xii
politics over life, 242–243
Politiques de l'amitié (Derrida), 90
Pompa, Leon, 23n14
Porter, Robert, 236–237
power: Esposito on, 81; the impolitical and, 181–182; Machiavelli on, 15–18
Preston, Paul, 77
Primitive Culture (Tylor), 113–114
Primo de Rivera, Pilar, 77
The Prince (Machiavelli), 15–18
privacy, 131
private property, 12–13, 49–51. *See also* debt

pro-life movement, 243–245
Proudhon, Pierre-Joseph, 49–50
Providence, 193–194
"El pudor de la historia" ("The Modesty of History") (Borges), 188–189

Quijano, Anibal, 174n16

racial discourse, 73
Raffoul, François, 53
Rancière, Jacques, 144, 153–155, 157–158
religion: Esposito on, 29, 30; Marx on, 29; Vico on, 20
resistance, 71, 110
rights and rights-based discourse, 57–58
Roman Catholicism and Political Form (Schmitt), 182–184
Roman law: the person and, 68–70, 74, 75, 103–104, 107, 127, 131, 145–146, 162–164, 213; on things, 167; things and, 165; Vico and, 5, 21
Rousseau, Jean-Jacques, 4–7, 10–15, 169, 226
Russell, Matheson, 160n30

Scheler, Max, 139n3
Schmitt, Carl, 182–184, 185, 196n10
Schopenhauer, Arthur, 72–73
Scorched (Mouawad): affirmative biopolitics and, 234–239; as catastrophe play and political play, 219–222; immunity paradoxes in, 220, 226–234; warring immunities and, 220
Searle, John, 139n3
The Second Treatise on Government (Locke), 51
security, 223–224
self-knowledge, 11–15
Serres, Michel, 172
sexuality, 136
S.H and Others v Austria, 250–251, 252
Simondon, Gilbert, 132–133, 164–165
"El simulacro" ("The Sham") (Borges), 190–191
Singer, Peter, 75, 139n3
Sloterdjik, Peter, 165
Social Contract (Rousseau), 14
Socialisme ou barbarie, 49
soggettività antagonista (antagonistic subjectivity), 47
Sources of the Self (Taylor), 136–137
sovereignty: Esposito on, 130; Hobbes on, 107
Spanish Civil War (1936-39), 77
Spencer, Herbert, 198n40
Spinoza, Baruch, 84, 164–165, 172
Stahl, Georg, 113
The State of Exception (Agamben), 48–49
Stein, Edith, 28–29, 41–42, 139n3
Stengers, Isabelle, 120, 171
Strathern, Marilyn, 116–117
surveillance, 131
Sykes-Picot Agreement (1916), 221

Tatián, Diego, 195n5
Tauber, Alfred, 35–36
Taylor, Charles, 136–137
tekhnè, 185–186, 189, 191
temporality, 11–13, 15–18
thanatopolitics: vs. affirmative biopolitics, 165; Esposito on, 225; Foucault on, 210; the impersonal and, 108; Nazi ideology and, 73–74, 79, 104, 117; neoliberalism and, 147–148; *Scorched* (Mouawad) and, 234–235
the third person: Benveniste on, xiv, 70, 78, 101–102

the third person—Esposito on: affirmative biopolitics and, 81–82, 151, 194; concept of, xiv, 70–71; genealogy of, 108–113, 132–134; International Relations and, 212–215; *munus* and, 143–144, 150–157, 158; Providence and, 194; theoretical grid for, 101–102; Weil and, 58, 76–77
Third Person (Esposito): on affirmative biopolitics, 93–95, 194; on human rights, 143; on the impersonal, 143–144, 150–157; on the living person, 101, 119; on neoliberalism, 143, 147–150; on the person, xiv, 127, 143, 144–147; on rights and rights-based discourse, 57–58
"tolerant immunity," 28–29, 35, 40–43
totalitarianism, 32–33, 41, 74, 134–136, 188–189
transformation masks, 118–119
Tronti, Mario, 88, 91–92, 93
Truth and Method (Gadamer), 7
Tuomela, Raimo, 139n3
Tylor, Edward Burnett, 113–114

Uexkuell, Jakob von, 84
The Unavowable Community (Blanchot), 49
United Nations (UN), 212–213
Universal Declaration of Human Rights (UDHR), 73–74, 106, 128, 213
the unthought, 110–111
Uriburu, José Félix, 197n26
L'uso dei corpi (*The Use of Bodies*) (Agamben), 91–92
utility, 13

verum, 6
Vibrant Matter (Bennett), 171
Vico, Giambattista: compared to Esposito and, 5–6; Croce on, 23n14, 24n20; Descartes and, 6; etymology and, 6; Gadamer on, 7; Giarrizzo on, 24n20; on Machiavelli, 15, 17–18; on origin of thought and society, 3–4; Pompa on, 23n14
Vico, Giambattista—Esposito on: bestial and social realms and, 3–5; as bourgeois thinker, 20–21; ideal eternal history and, 7–10; Machiavelli and, 4–7, 10, 15–20; the person and, 164–165; power and, 89; Rousseau and, 4–7, 10–15
Vico e Rousseau e il moderno Stato borghese (Esposito), 4
Villeneuve, Denis, 219
violence: Esposito on, 29–30; Machiavelli on, 43; Vico on, 10
Viriasova, Inna, 196n10
virtù, 17
Viveiros de Castro, Eduardo, 118, 123n22, 171–172, 174n16
voting rights, 127–128

war on terror, 27, 30–31
Weber, Max, 185
Weil, Simone: on community, xii; on the impersonal, xiv, 58, 76–77, 101–102, 193–194; on the person, 145, 151, 167
"What Is It Like to Be a Bat?" (Nagel), 172
Whitehead, Alfred North, 172
"Who is the Subject of the Rights of Man?" (Rancière), 153–155
Willerslev, Rane, 123n22
women, 77–78, 127–128, 154–155
World Association for Reproductive Medicine (WARM), 247–248

Yrigoyen, Hipólito, 197n26

zoé, 101, 115, 118, 129, 247